AF588149

Practical Neural Networks in Python and MATLAB

Chunwei Zhang · Tianpeng Li · Ying Dai · Li Sun ·
Ardashir Mohammadzadeh

Practical Neural Networks in Python and MATLAB

Chunwei Zhang
Shenyang University of Technology
Shenyang, China

Ying Dai
Northeastern University
Shenyang, China

Shenyang University of Technology
Shenyang, China

Ardashir Mohammadzadeh
Sakarya University
Sakarya, Türkiye

Shenyang University of Technology
Shenyang, China

Tianpeng Li
Shenyang University of Technology
Shenyang, China

Li Sun
Shenyang Jianzhu University
Shenyang, China

ISBN 978-3-032-14745-5 ISBN 978-3-032-14746-2 (eBook)
https://doi.org/10.1007/978-3-032-14746-2

This Springer imprint is published by the registered company Springer Nature Switzerland AG
The registered company address is: Gewerbestrasse 11, 6330 Cham, Switzerland

Contents

List of Figures

List of Tables

Chapter 1
Introduction

Abstract This chapter serves as an introduction to the book, providing a systematic review of the development of neural networks since the McCulloch-Pitts model was proposed in the 1940s, dividing it into three key phases. It highlights the core characteristics of neural networks, such as high fault tolerance, parallel processing capabilities, self-learning and self-adaptability, and powerful nonlinear approximation. The chapter extensively enumerates the broad applications of neural networks across various fields including image processing, signal analysis, pattern recognition, robotic control, healthcare, and economic forecasting, underscoring their utility as versatile tools. To establish the theoretical foundation for subsequent practical implementation, the chapter focuses on analyzing and comparing five fundamental training algorithms: Gradient Descent, Newton's Method, the Conjugate Gradient Method, the Quasi-Newton Method, and the Levenberg-Marquardt algorithm. Using flowcharts and mathematical formulations, it explains their respective principles, advantages, disadvantages, and suitable application scenarios, offering clear guidance for readers to select the appropriate optimization method based on specific problems.

1.1 Overview

The inception of neural network research dates to the 1940s with the seminal collaboration between psychologist Warren Sturgis McCulloch and mathematician Walter Pitts. They introduced the McCulloch-Pitts (MP) model, which laid the groundwork for subsequent neural network research. The evolution of neural networks can be charted through three distinct phases: the initial phase from 1947 to 1969, which was a period of prolific model and rule proposals, including the MP model, Hebb's learning rules, and the perceptron model; the transitional phase from 1970 to 1986, following a lull in research activity, which saw significant contributions such as John Hopfield's introduction of energy functions to networks, providing stability criteria and pathways for associative memory and optimal computation. In 1984, Geoffrey Hinton proposed the Boltzmann machine model, and in 1986, David Rumelhart, Geoffrey Hinton, and Ronald Williams introduced the error backpropagation (BP)

C. Zhang et al., *Practical Neural Networks in Python and MATLAB*,
https://doi.org/10.1007/978-3-032-14746-2_1

neural network, which has since become a cornerstone in the field. The current phase, from 1987 to the present, marks a period of global attention and research, propelling neural networks to new heights of development. Neural networks are characterized by several distinctive features:

1. High fault tolerance due to the distributed storage of information across network neurons.
2. Rapid calculations facilitated by parallel processing methods.
3. Self-learning, self-organization, and self-adaptability enable the network to handle uncertain or unknown systems.
4. The ability to approximate any complex non-linear relationships, with strong information synthesis capabilities, managing both quantitative and qualitative information, coordinating multiple input information relationships, and being well-suited for multi-information fusion and multimedia technology.

1.2 Some Applications of Neural Networks

Applications of neural networks span a wide range of fields:

1. Image processing, including edge detection, segmentation, compression, and recovery.
2. Signal processing, capable of handling communication, voice, ECG, and EEG signals, with applications in undersea sonar signal detection and classification for anti-submarine and mine-clearing operations.
3. Pattern recognition, successfully applied to handwritten characters, license plates, fingerprints, and voice recognition, as well as automatic target identification and positioning, robotic sensor image recognition, and seismic signal identification.
4. Robot control, coordinating the position of robotic eye-hand systems for troubleshooting and navigation of intelligent adaptive mobile robots.
5. Healthcare and medical applications, such as distinguishing between normal and abnormal heartbeats through multi-layer perceptron training, and BP network-based waveform classification and feature extraction in computer-aided clinical diagnosis.
6. Welding, where research has been conducted on parameter selection, quality inspection, prediction, and real-time control, with some results already in practical use.
7. Economics, with the capability to forecast short-term commodity and stock prices, as well as business creditworthiness.
8. Sensor signal processing, including correction of non-linear sensor output characteristics, fault detection, filtering, noise removal, compensation for environmental impacts, and multi-sensor information fusion.

9. The chemical field, where neural networks analyze pharmaceuticals, biochemistry, and chemical engineering, such as protein structure analysis, spectral analysis, and chemical reaction analysis.
10. Geography, with extensive applications in remote sensing image classification and the use of artificial neural network theory in GIS to enhance complex and comprehensive data analysis capabilities.
11. Additional applications are found in data mining, power systems, transportation, military, mining, agriculture, and meteorology.

1.3 Different Types of Neural Network Training

The most important training algorithms for several neural networks.

1.3.1 Gradient Descent

Regarding neural network training, one of the most critical algorithms is gradient descent, a simple yet effective first-order method that utilizes gradient vector information. The algorithm commences at an initial point and iteratively moves in the direction opposite to the gradient, adjusting weights until a termination condition is met. The iteration formula for gradient descent is given by:

$$w_{i+1} = w_i - d_i \cdot \eta_i, \quad i = 0, 1, ... \tag{1.1}$$

where, η is the learning rate, a parameter that can be fixed or dynamically updated through one-dimensional optimization methods. The following figure shows the flowchart of the gradient descent training method. As shown in the figure, the parameters are updated in two steps: the first step calculates the direction of gradient descent, and the second step calculates the appropriate learning rate.

The gradient descent method has a serious drawback in that it requires many iterative operations if the gradient change of the function exhibits a slender structure. Moreover, although the direction of gradient descent is the direction in which the value of the loss function decreases the fastest, this is not necessarily the fastest path to convergence. The following Fig. 1.1 depicts these problems.

The gradient descent method is our recommended algorithm when the neural network model is very large and contains thousands of parameters. Because this method only needs to store the gradient vector (n-space) and not the Hessian matrix (n2-space).

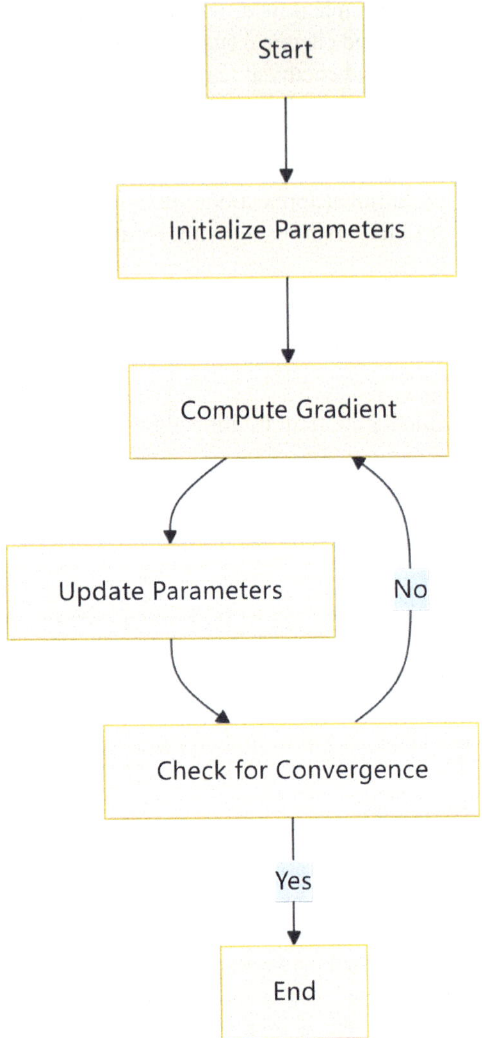

Fig. 1.1 The flowchart of the gradient descent training method

1.3.2 *Newton's Method*

Since Newton's algorithm uses the Hessian matrix, it is a second-order algorithm. The goal of this algorithm is to find a better learning direction using the second-order partial derivatives of the loss function.

We define $f\,(w_i) = f_i, \nabla f(w_i) = g_i \; and \; Hf(w_i) = H_i$. Estimate the function f at w_0 values using the Taylor expansion

$$f = f_0 + g_0 \cdot (w - w_0) + 0.5 \cdot (w - w_0)^2 \cdot H_0. \tag{1.2}$$

H_0 is the Hessian matrix value of the function f at w_0. At the minimum of $f(w)$ where $g = 0$, we obtain the second equation

$$g = g_0 + H_0 \cdot (w - w_0) = 0. \tag{1.3}$$

Therefore, initializing the parameters at w_0, the iterative formula for Newton's algorithm is

$$w_{i+1} = w_i - H_{i-1} \cdot g_i, \ i = 0, 1, ... \tag{1.4}$$

$H_{i-1} \cdot g_i$ is known as the Newton term. It is worth noting that if the Hessian matrix is a non-positive definite matrix, the parameters can move in the direction of the maximum value rather than the minimum value. Thus, the loss function value is not guaranteed to decrease at each iteration. To avoid this problem, we usually modify the equation of Newton's algorithm slightly:

$$w_{i+1} = w_i - (H_{i-1} \cdot g_i) \cdot \eta_i, \ i = 0, 1, ... \tag{1.5}$$

The learning rate η can either be set to a fixed value or dynamically adjusted. The vector $d = H_{i-1} \cdot g_i$ is referred to as the Newton training direction.

The following figure shows the flowchart of Newton's method. The updating of parameters is also divided into two steps: calculating the Newton training direction and the appropriate learning rate.

The performance of Newton's method is shown in Fig. 1.2. Starting from the same initial value to find the minimum of the loss function, it requires fewer steps than the gradient descent method.

However, the difficulty with Newton's method is the large amount of computational resources required to accurately compute the Hessian matrix and its inverse.

1.3.3 Conjugate Gradient Method (Conjugate Gradient)

The conjugate gradient method is between the gradient descent method and the Newton method. It was originally designed to solve the problem that the traditional gradient descent method converges too slowly. Unlike Newton's method, the conjugate gradient method also avoids computing and storing the Hessian matrix.

The conjugate gradient method searches along the conjugate directions and will usually converge faster than the gradient descent method. These training directions are conjugate to the Hessian matrix.

We define d as the training direction vector. Then, the parameter vector and the training direction training are initialized as w_0 and $d_0 = -g_0$, respectively, and the direction update formula for the conjugate gradient method is:

$$d_{i+1} = g_{i+1} + d_i \cdot \gamma_i, \ i = 0, 1, ... \tag{1.6}$$

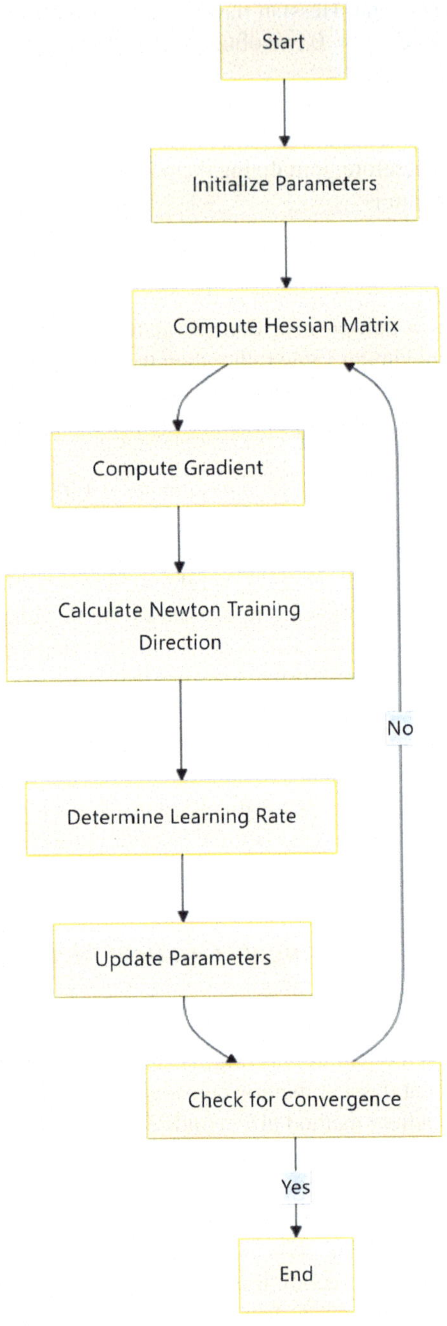

Fig. 1.2 The flowchart of Newton's method

where γ is the covariance parameter, and there are many ways to calculate it. Two of the commonly used methods were invented by Fletcher and Reeves and Polak and Ribiere, respectively. For all conjugate gradient algorithms, the training direction is periodically reset to a negative value of the gradient.

The update equation for the parameters is:

$$w_{i+1} = w_i - d_i \cdot g_i,\ i = 0, 1, ... \tag{1.7}$$

Figure 1.3 shows the flowchart of the training process of the conjugate gradient method. The step of parameter updating is divided into two steps: calculating the conjugate gradient direction and calculating the learning rate.

The efficiency of this method for training neural network models is better than the gradient descent method. Since the conjugate gradient method does not require the computation of the Hessian matrix, we also recommend it when the neural network model is large.

1.3.4 Quasi-Newton Method (Quasi-Newton Method)

Since Newton's method requires the computation of the Hessian matrix and its inverse matrix, which requires more computational resources, a variant algorithm called the Cauchy-Newton method has emerged to make up for the high computational effort. Instead of calculating the Hessian matrix and its inverse matrix directly, this method calculates the inverse matrix of the Hessian matrix at each iteration estimation, and only the first order partial derivatives of the loss function are used.

The Hessian matrix is composed of the second-order partial derivatives of the loss function. The main idea of the Cauchy-Newton method is to estimate the inverse matrix of the Hessian matrix using another matrix G, requiring only the first order partial derivatives of the loss function. The update equation for the Cauchy-Newton method can be written as:

$$w_{i+1} = w_i - (G_i \cdot g_i) \cdot \eta_i,\ i = 0, 1, ... \tag{1.8}$$

The learning rate η can either be set to a fixed value or dynamically adjusted. There are several different types of estimates of G for the inverse matrix of the Hessian matrix. Two commonly used types are the Davidon-Fletcher-Powell formula (DFP) and the Broyden-Fletcher-Goldfarb-Shanno formula (BFGS).

The flowchart of the Cauchy-Newton method is shown as Fig. 1.4. The steps of parameter updating are divided into calculating the Cauchy-Newton training direction and calculating the learning rate.

In many cases, this is the default algorithm of choice: it is faster than gradient descent and conjugate gradient methods without the need to accurately compute the Hessian matrix and its inverse.

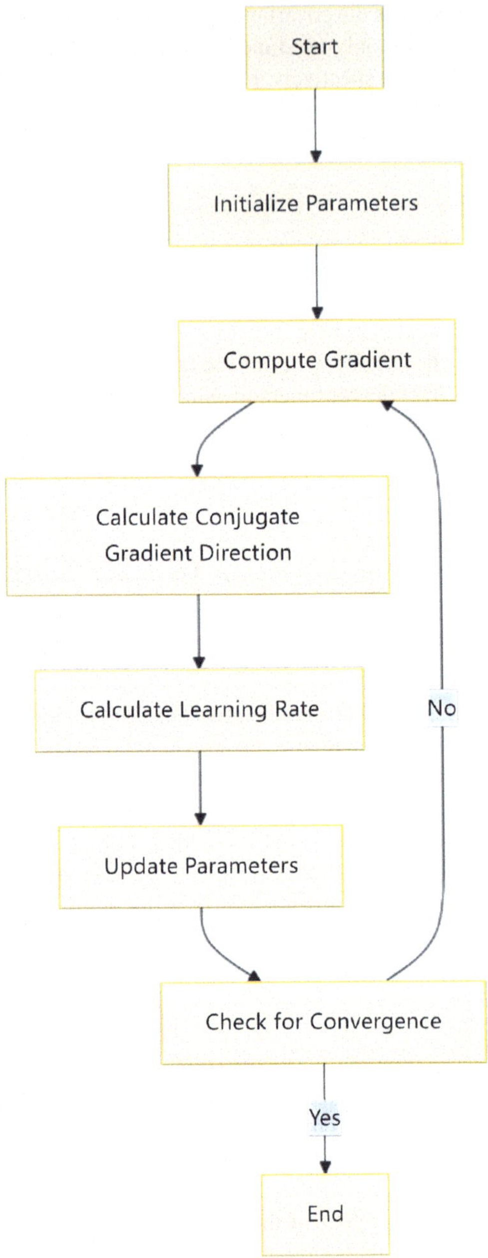

Fig. 1.3 The flowchart of the training process of the conjugate gradient method

Fig. 1.4 The flowchart of the Cauchy-Newton method

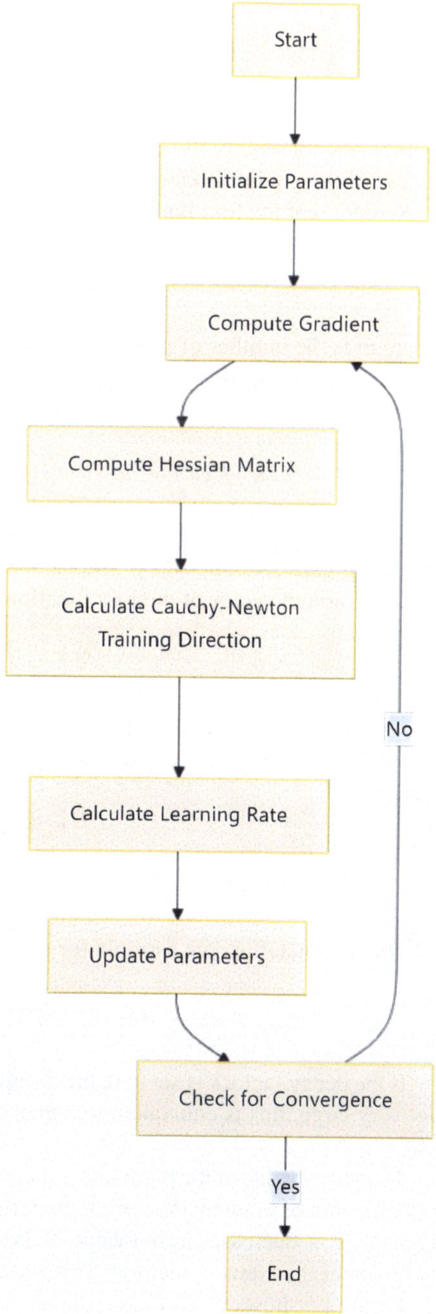

1.3.5 Levenberg-Marquardt Algorithm

The Levenberg-Marquardt algorithm, also known as the decaying least-squares method, targets the loss function, which is in the form of a sum-of-squares error. It also does not require the exact computation of the Hessian matrix, which requires the use of the gradient vector and the Jacobian matrix.

Assume that the loss function f is in the form of a sum-of-squares error:

$$f = \sum e_i^2,\ i = 1, \ldots, m, \tag{1.9}$$

where m is the number of training samples.

We define the Jacobian matrix of the loss function to consist of the partial derivatives of the error terms concerning the parameters, the

$$J_{i,j} f(w) = de_i / dw_j \ (i = 1, \ldots m \ \&\ j = 1, \ldots n), \tag{1.10}$$

m is the number of samples in the training set, and n is the number of parameters of the neural network. The size of the Jacobian matrix is m-n

The gradient vector of the loss function is:

$$\nabla f = 2J^T \cdot e, \tag{1.11}$$

e is the vector consisting of all error terms.

Finally, we can use this expression to estimate the computation of the Hessian matrix:

$$H_f \approx 2J^{T^{J+\lambda I}}, \tag{1.12}$$

λ is the decay factor to ensure that the Hessian matrix is positive, and I is the unit matrix.

The parameter update formula for this algorithm is as follows:

$$w_{i+1} = w_i - \left(J_i^T \cdot J_i + \lambda_i I\right)^{-1} \cdot \left(2J_i^T \cdot e_i\right),\ i = 0, 1, \ldots. \tag{1.13}$$

If the decay factorλ is set to 0, this is equivalent to being a Newton method. If λ is set very large, this is equivalent to a gradient descent method with a small learning rate.

The initial value of the parameter λ is very large, so the first few update steps are in the direction of gradient descent. If an iterative update fails at a step, λ expands a bit. Otherwise, λ decreases as the value of the loss decreases and Levenberg-Marquardt approaches the Newton method. This process speeds up the convergence.

Figure 1.5 shows the flowchart of the training process of the Levenberg-Marquardt algorithm. The first step is to calculate the loss value, gradient, and approximate Hessian matrix. Then decay parameters and decay coefficients.

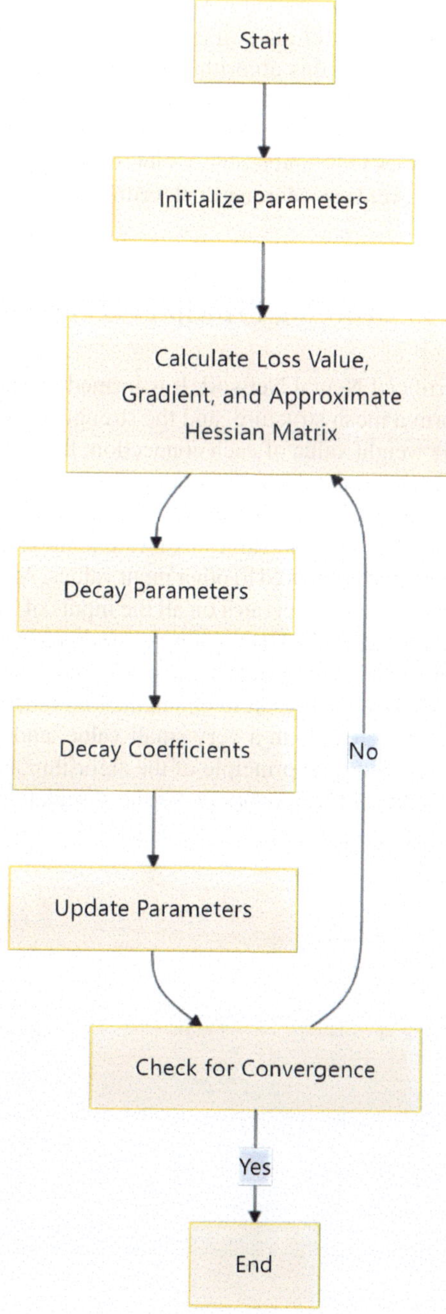

Fig. 1.5 The flowchart of the training process of the Levenberg-Marquardt algorithm

Since the Levenberg-Marquardt algorithm focuses on loss functions of the sum-of-squares error class, it is very fast in training neural network models for this type of error. But this algorithm also has some drawbacks. Firstly, it does not apply to other types of loss functions. Also, it is not compatible with regular terms. Finally, if the training data and the network model are very large, the Jacobian matrix also becomes large and requires a lot of memory. Therefore, we do not recommend using the Levenberg-Marquardt algorithm when the training data or the model is large.

1.4 Learning Principles of Neural Networks

Artificial Neural Network It is formed by simple neurons that are interconnected to form a mesh structure, and the strength of the connections is changed by adjusting the weight value of each connection, thus achieving perceptual judgment.

A neuron is the basic computational unit of a neural network, also known as a node or unit. It can accept inputs from other neurons or external data and then compute an output. Each input value has a weight, the size of which depends on the importance of that input compared to other input values. A specific function f is then executed on the neuron, which operates on all the inputs of the neuron and their weights. Traditional neural networks have a simple structure, where the input parameters are randomly initialized during training and a loop is started to compute the output, compare it with the actual result to obtain the loss function, update the variables so that the loss function results in a very small value, and stop the loop when the error threshold is reached. The principle of the algorithm is to continuously change the connection weights of the network under the stimulation of external input samples.

Chapter 2
Multilayer Perceptron (MLP) Neural Networks

Abstract This chapter provides a comprehensive exploration of Multilayer Perceptron (MLP) Neural Networks, beginning with the fundamental architecture of neurons and progressing to complex network structures. It details the error backpropagation algorithm as the cornerstone of MLP training, explaining both forward and backward phases of weight adjustment. The chapter demonstrates practical implementation through extensive Python code examples, covering neural network initialization, forward propagation, and gradient descent-based training. Significant attention is given to MLP applications in classification tasks, particularly in medical diagnostics using diabetes datasets, with comparisons of various machine learning algorithms. The chapter also addresses critical training considerations including over-parameterization and over-training, providing practical strategies for optimal network configuration. Throughout, the content maintains a strong practical focus with executable code snippets for system estimation, classification, and performance evaluation.

A neuron is the most basic unit of information processing in a neural network, which is capable of handling both linear and nonlinear data. The neuron model consists of three elements: synapses, adders, and activation functions. In addition, it contains external bias, which, depending on the sign and magnitude of its value, increases or decreases the input signal weighting and input to the neural network through the activation function.

Neural Networks

Neural networks are characterized by a layered structure and can be divided into three categories depending on their function: input, hidden, and output layers. Neurons are called source nodes, hidden nodes, and output nodes according to the layer of the network in which they are located. According to the different information flow in the network, neural network construction structure can be divided into.

Feedforward Networks and Recursive Networks. In the feed-forward network structure, information is passed backward from the source node step by step until the final output node, while in the recursive network structure, the output of the neuron and its input will form a feedback loop. The way a neural network is constructed is closely linked to the learning algorithm used to train the network.

C. Zhang et al., *Practical Neural Networks in Python and MATLAB*,
https://doi.org/10.1007/978-3-032-14746-2_2

2.1 Training Based on Error Backpropagation

Multilayer Neural Networks and Backpropagation Learning Algorithms The main feature of backpropagation neural networks is that the outputs of neurons are fed back to their input neurons. The backpropagation algorithm is based on multilayer perceptron (neural network), is a popular algorithm for training multilayer perceptron, and consists of two phases, "forward phase" and "reverse phase":

1. Forward phase: The weights and biases of each synapse in the network are fixed, and the input signal (x_i) propagates along the synapses in the network from the input layer until it reaches the output layer.
2. Reverse phase: the error signal (e_j) between the output signal (y_j) and the desired value (d_i) of the forward phase is propagated from the output layer to the input layer, and the synaptic weights w_{kj} and bias of the neurons it passes through are modified b_k.

The signal back propagation algorithm is based on the forward propagated signal flow of neurons under the action of Least Mean Square (LMS: Least Mean Square) error.

2.2 Implementation with Python

Neural Network Toolkit for Python

Given the inherent nature of neural networks, which often requires handling a vast array of training data, intricate arithmetic operations, and complex programming tasks, Python, a versatile programming language, offers a robust ecosystem for neural network development. Python provides numerous libraries, such as TensorFlow, Keras, and PyTorch, which are equipped with a wide array of functions and commands to facilitate neural network tasks.

These libraries serve as exceptional platforms for neural network training and simulation. They are frequently employed for initializing, simulating, designing, tuning, and optimizing neural networks. The integrated processing capabilities, user-friendly interfaces, and visual demonstration of the training process significantly reduce the amount of time spent on programming, allowing individuals without extensive expertise in neural networks to effectively utilize them.

Python, a high-level programming language known for its readability and efficiency, is widely used in the fields of data science, artificial intelligence, and engineering. It supports powerful data structures and provides a simple yet elegant syntax for matrix operations, which is crucial for neural network computations. The availability of these tools and libraries in Python has made it a preferred choice for professionals and enthusiasts alike in the realm of neural network applications.

2.3 Application of Neural Networks in Classification

Multi-layer perceptron is the promotion of single-layer perceptron, which can successfully solve the non-linear differentiable problems that cannot be solved by single-layer perceptron, and introduce a hidden layer between the input layer and the output layer as the "internal representation" of the input pattern, so that the single-layer perceptron can be turned into a multi-layer perception.

With the rapid development of science and technology, artificial intelligence has gradually become a research hotspot, in which the field of computer vision is particularly compelling. Image classification, as one of the core tasks of computer vision, aims to allow computers to automatically recognize and classify the contents of images. In recent years, Convolutional Neural Network (CNN), as a deep learning model, has achieved remarkable results in the image classification task and become the mainstream technology in this field. Convolutional neural network has a powerful feature extraction capability, which can automatically learn the hierarchical representation of images and effectively recognize complex patterns and structures in images. Compared with traditional image processing methods, CNNs show higher accuracy and robustness in processing large-scale image data.

The image classification problem is a core task in the field of computer vision, which involves automatically analyzing and understanding the input image to identify the objects or scenes contained in the image and classify them into predefined categories. With the rapid development of the digital era, image data are being used more and more widely in various fields, such as medical diagnosis, security monitoring, intelligent transportation, human-computer interaction, and so on. The study of the image classification problem has important practical application value and social significance.

The image classification problem is an important computer vision task, which is of great significance in promoting the development and application of artificial intelligence technology. And convolutional neural network, as an effective image classification method, plays an increasingly important role in practical applications. The research on the application of convolutional neural networks in image classification has important theoretical value and practical application significance. Convolutional Neural Network (CNN) has become a very effective method in image classification. CNN can extract high-level feature representations from the original pixel data by simulating the connection patterns between neurons in the human visual system, thus realizing the effective classification of images. Compared with traditional image classification methods, CNNs have stronger feature learning abilities and higher classification accuracy and, thus, have achieved remarkable results in many practical applications.

Development of Convolutional Neural Networks (CNNs) and Their Application in Image Classification Background Convolutional Neural Networks (CNNs), as a kind of deep learning model, have been closely linked to the advancement of image classification techniques. CNNs were initially inspired by the biological visual system, in particular, Hubel and Wiesel's study of the cat's visual cortex, in which they found simple and complex cells that respond selectively to the spatial localization and orientation of visual stimuli. This biological discovery facilitated the initial design of CNNs, allowing neural networks to process image data efficiently. The development of these network architectures has not only improved the accuracy of image classification but also promoted the application of CNNs in a wider range of fields, such as object detection, image segmentation, video analytics, etc. The background of CNNs in image classification, from the initial recognition of simple characters to today's accurate recognition in complex scenarios, exemplifies the tremendous progress of AI technology in mimicking the human visual perceptual ability.

2.4 MLP Neural Networks

The simplest type of neuron modeling is the perceptron. Since analyzing several perceptron types in different layers is difficult, we will begin by examining one perception.

According to Fig. 2.1, a perceptron consists of a series of external inputs, an internal input called a bias, a threshold, and an output.

Perceptron learning is the process of determining the appropriate values of w. As a result, perceptron learning is defined as the sum of all possible real weight vector values. The following equation determines the output of a perception:

$$\begin{cases} 1 \; w_0 + x_1 w_1 + \cdots + x_n w_n > 0 \\ 0 \; w_0 + x_1 w_1 + \cdots + x_n w_n \leq 0. \end{cases} \tag{2.1}$$

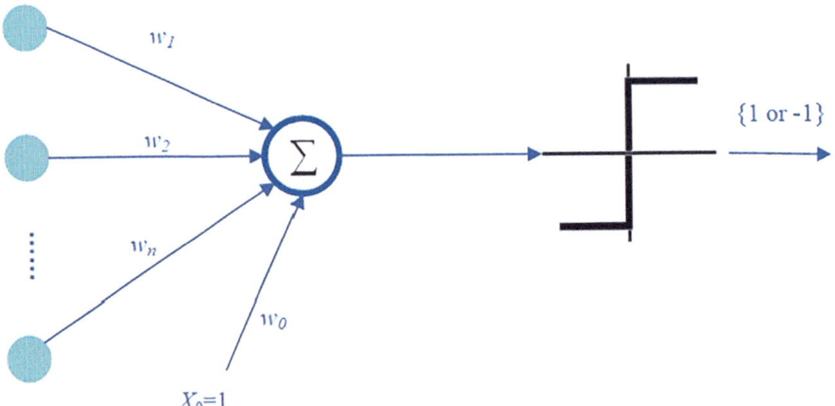

Fig. 2.1 The model of a perceptron

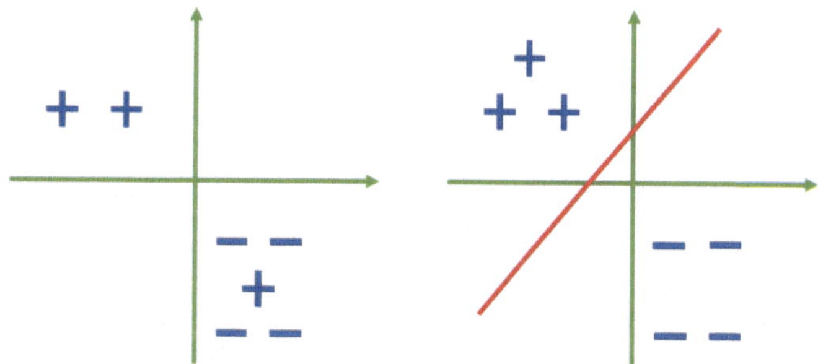

Fig. 2.2 Example of separation

A perceptron can only learn examples that can be separated linearly. These are instances that can be completely separated by a hyperplane. Consider the following Fig. 2.2.

Example: OR Function

A perceptron is capable of dividing a space into two parts. Thus, only functions with positive and negative outputs that can be divided into two parts in space are accurately obtained from a perceptron. We managed to train a perceptron in an OR function, as shown in the Fig. 2.3.

The layers of a multilayer network are as follows:

1. Input Layer: It receives data fed into the network in its raw form.
2. Hidden Layers: Inputs and the weight of connections between them and the hidden layers determine the performance of these layers. When a hidden unit should be activated, the weights between input and hidden units are employed to determine when it should be activated.

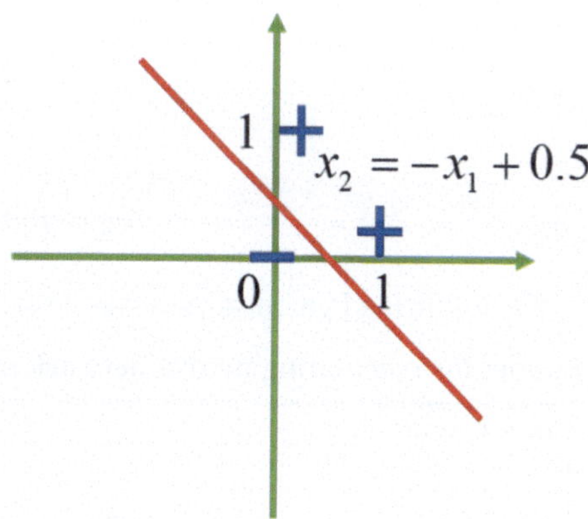

Fig. 2.3 Second example of separation

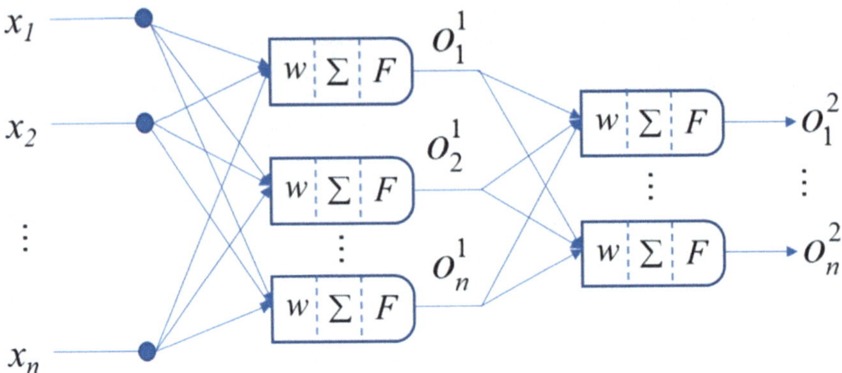

Fig. 2.4 A neural network with three layers

3. Output Layer: The performance of the output unit is determined by the activity of a hidden layer and the weights of connections between the hidden and output layers.

Figure 2.4 depicts a three-layer neural network. Where, $\left[x_1, ..., x_{n_0}\right]$ denotes the neural network input vectors, and $\left[o_1^1, ..., o_h^1\right]$ indicates the middle layer outputs, whereas $\left[o_1^2, ..., o_n^2\right]$ represents the neural network outputs.

If the matrix of weights connected to the middle and input layer neurons is represented by w_1 and the matrix of weights connected to the middle and output layer neurons is represented by w_2, the input vector is represented by x, the activation functions in the middle layer and the output are represented by F_1 and F_2, the output vector of the middle layer is represented by o_1, and the output network vector is represented by o_1, the neural network output is as follows:

$$net_1 = w_1^T x, \tag{2.2}$$

$$o_1 = F_1\left(net_1\right), \tag{2.3}$$

$$net_2 = w_2^T o_1, \tag{2.4}$$

$$o_2 = F_2\left(net_2\right). \tag{2.5}$$

The following Python snippet was written to calculate the neural network output:

Snippet for calculating neural network output

```
import numpy as np

# Define the sigmoid activation function and its derivative
def sigmoid(x):
    return 1 / (1 + np.exp(-x))

def sigmoid_derivative(x):
```

```
    return x * (1 - x)

# Initialize the weights and biases
input_size = 3  # Number of neurons in the input layer
hidden_size = 2  # Number of neurons in the hidden layer
output_size = 1  # Number of neurons in the output layer

# Randomly initialize weights and biases
W1 = np.random.rand(input_size, hidden_size)
b1 = np.random.rand(hidden_size)
W2 = np.random.rand(hidden_size, output_size)
b2 = np.random.rand(output_size)

# Input data
X = np.array([[0.5, -0.2, 0.1]])  # Features of a single sample

# Forward propagation
Z1 = np.dot(X, W1) + b1  # Linear combination of inputs and weights plus
    bias
A1 = sigmoid(Z1)  # Activation of the hidden layer
Z2 = np.dot(A1, W2) + b2  # Linear combination of hidden layer activations
    and weights plus bias
A2 = sigmoid(Z2)  # Activation of the output layer

# Output the result of the neural network
print("Neural Network Output:", A2)
```

2.5 Training Based on Error Backpropagation

This method, introduced by Rumelhart and McClelland in 1986, is utilized by feedforward neural networks. The term "feedforward" indicates the arrangement of artificial neurons in sequential layers that pass their outputs (signals) in a forward direction. The backpropagation process involves sending error signals backward through the network to adjust the weights, followed by reprocessing the input from the input layer to the output layer. This supervised learning technique requires that input samples are tagged with their corresponding expected outputs, which are known beforehand. The network's output is then compared against these target outputs to calculate the error. The algorithm operates under the premise that network weights are initially chosen randomly. During each iteration, the network's output is determined, and the weights are adjusted based on the discrepancy between the actual and desired outputs, aiming to reduce the error over time.

The accompanying diagram illustrates the learning process of a basic neural network. An adaptive formula for the synaptic weights is derived by finding the optimal value of d, ensuring that the network's output closely matches d or that the neural network learns the optimal response d. A backpropagation solution that employs gradient descent based on the chain rule can address this challenge. The specific steps of this method are as follows: The squared instantaneous error between the target value d and the network's output at step k is considered the standard function for evaluating network performance (Figs. 2.5 and 2.6).

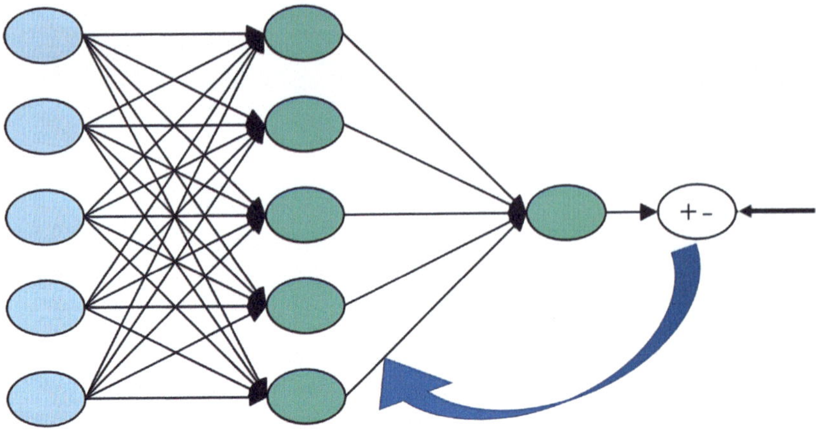

Fig. 2.5 A simple neural network with one output

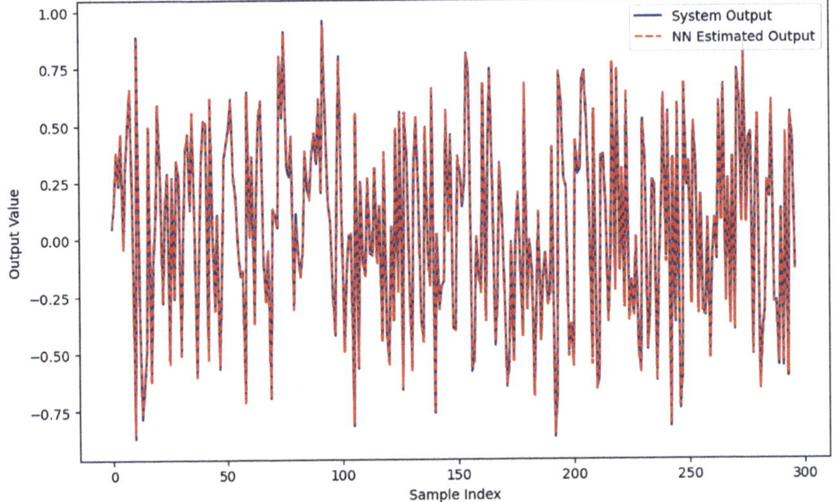

Fig. 2.6 System output versus NN estimated output

$$E(k) = \frac{1}{2}e^2(k) = \frac{1}{2}(d\,(k) - o_2\,(k))^2, \tag{2.6}$$

$$w_2\,(k+1) = w_2\,(k) + \left(-\eta_w \cdot \frac{\partial E\,(k)}{\partial w_2\,(k)}\right), \tag{2.7}$$

$$\frac{\partial E(k)}{\partial w_2(k)} = \frac{\partial E(k)}{\partial o_2(k)} \frac{\partial o_2(k)}{\partial net_2(k)} \frac{\partial net_2(k)}{\partial w_2(k)} = -eF_2'o_1, \tag{2.8}$$

$$\frac{\partial E(k)}{\partial w_1^j(k)} = \frac{\partial E(k)}{\partial e(k)}\frac{\partial e(k)}{\partial o_2(k)}\frac{\partial o_2(k)}{\partial net_2(k)}\frac{\partial net_2(k)}{\partial o_1^j(k)}\frac{\partial o_1^j(k)}{\partial net_1^j(k)}\frac{\partial net_1^j(k)}{\partial w_1^j(k)} = -eF'_2 w_2^j F'_1 x. \tag{2.9}$$

In the context of the error backpropagation algorithm, the symbol η_w denotes the learning rate associated with gradient descent. The weight connecting the j-th neuron to the output is represented by w_1^j, and the vector of weights associated with the j-th neuron in the first layer is denoted by w_j. The activation function derivatives for the output layer and the first hidden layer are indicated by $\frac{\partial f}{\partial z}$ and $\frac{\partial f}{\partial z_1}$, respectively. The training process involves iteratively applying these adjustments until the cost function, which measures the network's error, is minimized to an acceptable level.

2.6 Implementation in Python

Example:

Use a two-layer MLP neural network to estimate the following function.

$$y_p(k+1) = f\left(y_p(k),\ y_p(k-1),\ y_p(k-2),\ u(k),\ u(k-1)\right), \tag{2.10}$$

where

$$f(x_1,\ x_2,\ x_3,\ x_4,\ x_5) = \frac{x_1 x_2 x_3 x_5 (x_3 - 1) + x_4}{1 + x_2^2 + x_3^3}$$

To train a neural network with an arbitrary number of inputs and outputs, the following function was coded in Python:

MLP neural network training function with gradient descent.

```
import numpy as np

# Define the derivative of the bipolar sigmoid activation function
def df1(x):
    return 2 * np.exp(-x) / (1 + np.exp(-x))**2

def df2(x):
    return 2 * np.exp(-x) / (1 + np.exp(-x))**2

def GD_MLP(e, eta, u, o1, net1, net2, w1, w2):
    # Perform a single iteration of gradient descent for a multi-layer
        perceptron
    delta2 = e * df2(net2)  # Compute delta for the second layer
    w2 = w2 + eta * o1 * delta2[:, np.newaxis]  # Update weights for the
        second layer

    delta1 = (w2 * delta2) * df1(net1)  # Compute delta for the first layer
    w1 = w1 + eta * u * delta1  # Update weights for the first layer

    return w1, w2

# Example usage
```

```
# Sample input values, replace with actual values as needed
e = np.array([1.0, 2.0])  # Error vector
eta = 0.1  # Learning rate
u = np.array([1.0, 2.0])  # Input vector
o1 = np.array([1.0, 2.0])  # Output of the first layer
net1 = np.array([1.0, 2.0])  # Activation of the first layer
net2 = np.array([1.0, 2.0])  # Activation of the second layer
w1 = np.array([[1.0, 2.0], [3.0, 4.0]])  # Weights of the first layer
w2 = np.array([1.0, 2.0])  # Weights of the second layer

# Call the function
w1, w2 = GD_MLP(e, eta, u, o1, net1, net2, w1, w2)
print("Updated w1:", w1)
print("Updated w2:", w2)
```

Nonlinear system estimation snippet using neural networks.

```
import numpy as np
import matplotlib.pyplot as plt

# Define the feedforward neural network function with Leaky ReLU activation
def FeedForward_NN(x, b1, b2, w1, w2):
    net1 = np.dot(w1.T, x) + b1
    o1 = np.where(net1 > 0, net1, 0.01 * net1)  # Leaky ReLU activation
    net2 = np.dot(w2.T, np.vstack((b2, o1)))
    o2 = net2  # Linear output
    return o1, o2, net1, net2

# Define the gradient descent function with L2 regularization and gradient
    clipping
def GD_MLP(e, eta, x, o1, net1, net2, w1, w2, lambda_reg, clip_value=10.0):
    delta2 = np.clip(e, -clip_value, clip_value)
    w2 += eta * (np.dot(np.vstack((b2, o1)), delta2) - lambda_reg * w2)

    delta1 = delta2 * w2[1:] * (o1 > 0)
    delta1 = np.clip(delta1, -clip_value, clip_value)
    w1 += eta * (np.dot(x, delta1.T) - lambda_reg * w1)
    return w1, w2

# Generate training and testing data
u = 2 * (np.random.rand(1000, 1) - 0.5)
y = np.zeros((1000, 1))
for t in range(3, len(u)):
    y[t] = (y[t-1] * y[t-2] * y[t-3] * u[t-2] * (y[t-3] - 1) + u[t-1]) / (1
        + y[t-2]**2 + y[t-3]**2)
    y[t] = np.clip(y[t], -1, 1)

# Input data
data = np.column_stack((np.zeros((997, 1)), y[1:-2], np.zeros((997, 1)), y
    [2:-1], np.zeros((997, 1)), y[3:], np.zeros((997, 1)), u[1:-2], np.
    zeros((997, 1)), u[2:-1]))

# Shuffle and split data
nn = np.random.permutation(997)
Train_data = data[nn[:700], :]
Target_train = y[3:][nn[:700]]

Test_data = data[nn[700:], :]
Target_test = y[3:][nn[700:]]

# Normalize input data
def normalize_data(data):
    max_values = np.max(np.abs(data), axis=0)
```

```
    max_values[max_values == 0] = 1  # Avoid division by zero
    return data / max_values

Train_data = normalize_data(Train_data)
Test_data = normalize_data(Test_data)

# Neural network initial parameters
num_neuron = 4
w1 = np.random.randn(10, num_neuron) * 0.01
w2 = np.random.randn(num_neuron + 1, 1) * 0.01
eta = 0.01  # Further reduced learning rate
b1 = np.zeros((num_neuron, 1))
b2 = 0
lambda_reg = 0.0001

# Training
MSE_Train = np.zeros(100)
MSE_Test = np.zeros(100)
for epoch in range(1, 101):
    print(f'Epoch: {epoch}')
    E = 0
    nn = np.random.permutation(700)
    for ii in nn:
        x = Train_data[ii, :].reshape(-1, 1)
        o1, o2, net1, net2 = FeedForward_NN(x, b1, b2, w1, w2)
        e = Target_train[ii] - o2
        w1, w2 = GD_MLP(e, eta, x, o1, net1, net2, w1, w2, lambda_reg)
        E += (e**2).item()
    MSE_Train[epoch-1] = E / 700
    print(f'Train MSE: {MSE_Train[epoch-1]}')

# Testing and collect predictions
y_pred = []
E = 0
for ii in range(297):
    x = Test_data[ii, :].reshape(-1, 1)
    o1, o2, net1, net2 = FeedForward_NN(x, b1, b2, w1, w2)
    e = Target_test[ii] - o2
    E += (e**2).item()
    y_pred.append(o2.item())  # Append the predicted value

MSE_Test[-1] = E / 297
print(f'Test MSE: {MSE_Test[-1]}')

# Plot system output and neural network estimated output
plt.figure(figsize=(10, 6))
plt.plot(Target_test, label='System Output', color='blue')
plt.plot(y_pred, label='NN Estimated Output', color='red', linestyle='--')
plt.title('System Output vs NN Estimated Output')
plt.xlabel('Sample Index')
plt.ylabel('Output Value')
plt.legend(loc='upper right')
plt.show()
```

2.7 Application of Neural Network in Classification

Neural networks play a crucial role in medical diagnostics, particularly in the classification of diabetes. This study aims to determine the most effective classification algorithm for diabetes diagnosis by comparing several widely used techniques. Data are sourced from a comprehensive medical database, which includes various patient attributes such as age, BMI, glucose levels, and other relevant clinical indicators. You can download the data of the third tutorial session at the website.[1]

Python classification script with the desired number of inputs and outputs

Exploring the data

```
import pandas as pd
df = pd.read_csv('diabetes.csv')
df.head(10)

df.info()

# Importing essential libraries
import matplotlib.pyplot as plt
import seaborn as sns
import numpy as np
%matplotlib inline

# Create the plot
plt.figure(figsize=(10,7))
ax = sns.countplot(x='Outcome', data=df, palette='viridis')

# Removing the unwanted spines
ax.spines['top'].set_visible(False)
ax.spines['right'].set_visible(False)

# Headings
plt.xlabel('Has Diabetes')
plt.ylabel('Count')

# Annotate each bar with the count
for p in ax.patches:
        ax.annotate(f'{p.get_height()}', (p.get_x() + p.get_width() / 2., p.
            get_height()),
                    ha='center', va='center', fontsize=11, color='gray', xytext
                    =(0, 10),
                    textcoords='offset points')

plt.show()

# Replacing the 0 values from ['Glucose','BloodPressure','SkinThickness','
    Insulin','BMI'] by NaN
df_copy = df.copy(deep=True)
df_copy[['Glucose','BloodPressure','SkinThickness','Insulin','BMI']] =
    df_copy[['Glucose','BloodPressure','SkinThickness','Insulin','BMI']].
    replace(0,np.NaN)
df_copy.isnull().sum()

```

[1] http://www.

```
# Plotting histogram of dataset before replacing NaN values
p = df_copy.hist(figsize = (15,15))

# Replacing NaN value by mean, median depending upon distribution
df_copy['Glucose'] = df_copy['Glucose'].fillna(df_copy['Glucose'].mean())
df_copy['BloodPressure'] = df_copy['BloodPressure'].fillna(df_copy['
    BloodPressure'].mean())
df_copy['SkinThickness'] = df_copy['SkinThickness'].fillna(df_copy['
    SkinThickness'].median())
df_copy['Insulin'] = df_copy['Insulin'].fillna(df_copy['Insulin'].median())
df_copy['BMI'] = df_copy['BMI'].fillna(df_copy['BMI'].median())

# Plotting histogram of dataset after replacing NaN values
p = df_copy.hist(figsize=(15,15))

from sklearn.model_selection import train_test_split
from sklearn.metrics import accuracy_score, classification_report

X = df_copy.drop(columns='Outcome')
y = df_copy['Outcome']

X_train, X_test, y_train, y_test = train_test_split(X, y, test_size=0.20,
    random_state=0)
print('X_train size: {}, X_test size: {}'.format(X_train.shape, X_test.shape
    ))

# Feature Scaling
from sklearn.preprocessing import StandardScaler
sc = StandardScaler()
X_train = sc.fit_transform(X_train)
X_test = sc.transform(X_test)
results = []
```

K-nearest neighbors

```
from sklearn.neighbors import KNeighborsClassifier

# Create an instance of the KNeighbors classifier
knn = KNeighborsClassifier(n_neighbors=3)

# Model training
knn.fit(X_train, y_train)

# Prediction
y_pred = knn.predict(X_test)

# Accuracy evaluation
accuracy_knn = accuracy_score(y_test, y_pred)
results.append(accuracy_knn)

# Classification Report
print(classification_report(y_test, y_pred))
```

Naive Bayes

```
from sklearn.naive_bayes import GaussianNB

# Create an instance of the Native Bayes classifier
gnb = GaussianNB()

# Model training
gnb.fit(X_train, y_train)
```

```

# Prediction
y_pred = gnb.predict(X_test)

# Accuracy evaluation
accuracy_gnb = accuracy_score(y_test, y_pred)
results.append(accuracy_gnb)

# Classification Report
print(classification_report(y_test, y_pred))
```

SVM

```
from sklearn.svm import SVC

# Create an instance of the SVM
svm = SVC(kernel='linear', C=1.0, random_state=42)

# Model training
svm.fit(X_train, y_train)

# Prediction
y_pred = svm.predict(X_test)

# Accuracy evaluation
accuracy_svm = accuracy_score(y_test, y_pred)
results.append(accuracy_svm)

# Classification Report
print(classification_report(y_test, y_pred))
```

Decision tree

```
from sklearn.tree import DecisionTreeClassifier

# Create an instance of the Decision Tree
dt = DecisionTreeClassifier(random_state=42)

# Model training
dt.fit(X_train, y_train)

# Prediction
y_pred = dt.predict(X_test)

# Accuracy evaluation
accuracy_dt = accuracy_score(y_test, y_pred)
results.append(accuracy_dt)

# Classification Report
print(classification_report(y_test, y_pred))
```

Random forest

```
from sklearn.ensemble import RandomForestClassifier

# Create an instance of the Random forest
rf = RandomForestClassifier(n_estimators=100, random_state=42)

# Model training
rf.fit(X_train, y_train)

# Prediction
```

```
y_pred = rf.predict(X_test)

# Accuracy evaluation
accuracy_rf = accuracy_score(y_test, y_pred)
results.append(accuracy_rf)

# Classification Report
print(classification_report(y_test, y_pred))
```

Logistic regression

```
from sklearn.linear_model import LogisticRegression

# Create an instance of the Logistic Regression
log_reg = LogisticRegression(random_state=42)

# Model training
log_reg.fit(X_train, y_train)

# Prediction
y_pred = log_reg.predict(X_test)

# Accuracy evaluation
accuracy_log_reg = accuracy_score(y_test, y_pred)
results.append(accuracy_log_reg)

# Classification Report
print(classification_report(y_test, y_pred))

results

res = pd.DataFrame([results], columns=['K-NN', 'Naive Bayes', 'SVM', '
    Decision Tree', 'Random Forest', 'Logistic Regression'])
res.head()

# Transpose the DataFrame for ease of plotting
res = res.T
res.columns = ['Accuracy']

# Creating a bar chart
plt.figure(figsize=(10, 6))  # Set the chart size
bars = res['Accuracy'].plot(kind='bar', color='skyblue')
plt.title('Accuracy of Classification Models')  # Title of the chart
plt.xlabel('Model')
plt.ylabel('Accuracy')
plt.xticks(rotation=45)
plt.ylim(0., 1.0)  # Set limits for the Y axis for better readability

# Add value labels above the columns
for bar in bars.patches:
    plt.annotate(format(bar.get_height(), '.4f'),
                 (bar.get_x() + bar.get_width() / 2,
                  bar.get_height()), ha='center', va='center',
                 size=10, xytext=(0, 8),
                 textcoords='offset points')
plt.show()
```

2.8 Over-Parameterization

Optimizing the architecture of neural networks, particularly in terms of neuron count and structure, is a crucial aspect of training. Essentially, the quantity of tunable parameters within a neural network should align with the volume of training data to ensure effective learning. Conversely, over-parameterization arises when the network contains excessive parameters relative to the available training data, leading to suboptimal training outcomes.

In a prior experiment conducted using Python, the script was executed with varying numbers of neurons in the hidden layer to illustrate this issue. Over-parameterization is evident when the mean squared error (MSE) curves for training and testing data diverge, and the MSE for the test data begins to rise. This behavior highlights the importance of balancing network complexity with the amount of training data to achieve optimal performance.

Python script for detecting the optimal number of middle layer neurons to avoid over- parameterization.

```
import pandas as pd
import numpy as np
import matplotlib.pyplot as plt
import seaborn as sns
%matplotlib inline

# Load the dataset
df = pd.read_csv('diabetes.csv')
df.head(10)

df.info()

# Create the plot
plt.figure(figsize=(10,7))
ax = sns.countplot(x='Outcome', data=df, palette='viridis')

# Removing the unwanted spines
ax.spines['top'].set_visible(False)
ax.spines['right'].set_visible(False)

# Headings
plt.xlabel('Has Diabetes')
plt.ylabel('Count')

# Annotate each bar with the count
for p in ax.patches:
    ax.annotate(f'{p.get_height()}', (p.get_x() + p.get_width() / 2., p.
        get_height()),
                ha='center', va='center', fontsize=11, color='gray', xytext
                    =(0, 10),
                textcoords='offset points')

plt.show()

# Replacing the 0 values from ['Glucose','BloodPressure','SkinThickness','
    Insulin','BMI'] by NaN
df_copy = df.copy(deep=True)
```

```
df_copy[['Glucose','BloodPressure','SkinThickness','Insulin','BMI']] =
    df_copy[['Glucose','BloodPressure','SkinThickness','Insulin','BMI']].
    replace(0,np.NaN)
df_copy.isnull().sum()

# Plotting histogram of dataset before replacing NaN values
p = df_copy.hist(figsize = (15,15))

# Replacing NaN value by mean, median depending upon distribution
df_copy['Glucose'] = df_copy['Glucose'].fillna(df_copy['Glucose'].mean())
df_copy['BloodPressure'] = df_copy['BloodPressure'].fillna(df_copy['
    BloodPressure'].mean())
df_copy['SkinThickness'] = df_copy['SkinThickness'].fillna(df_copy['
    SkinThickness'].median())
df_copy['Insulin'] = df_copy['Insulin'].fillna(df_copy['Insulin'].median())
df_copy['BMI'] = df_copy['BMI'].fillna(df_copy['BMI'].median())

# Plotting histogram of dataset after replacing NaN values
p = df_copy.hist(figsize=(15,15))

from sklearn.model_selection import train_test_split
from sklearn.preprocessing import StandardScaler
from sklearn.metrics import accuracy_score, classification_report

# Prepare the data
X = df_copy.drop(columns='Outcome')
y = df_copy['Outcome']

# Split the data
X_train, X_test, y_train, y_test = train_test_split(X, y, test_size=0.20,
    random_state=0)
print('X_train size: {}, X_test size: {}'.format(X_train.shape, X_test.shape
    ))

# Feature Scaling
sc = StandardScaler()
X_train = sc.fit_transform(X_train)
X_test = sc.transform(X_test)

# Importing necessary libraries for the neural network
from sklearn.neural_network import MLPClassifier

# Create an instance of the MLPClassifier with a large number of neurons and
    layers
# This is an example of over-parameterization
mlp = MLPClassifier(hidden_layer_sizes=(100, 100, 100, 100, 100), max_iter
    =1000, random_state=42)

# Model training
mlp.fit(X_train, y_train)

# Prediction
y_pred = mlp.predict(X_test)

# Accuracy evaluation
accuracy_mlp = accuracy_score(y_test, y_pred)
print(f"Accuracy of MLP: {accuracy_mlp}")

# Classification Report
print(classification_report(y_test, y_pred))

# Plotting the training and validation loss to show signs of overfitting
plt.figure(figsize=(10, 6))
```

```
plt.plot(mlp.loss_curve_, label='Training Loss')
plt.xlabel('Iteration')
plt.ylabel('Loss')
plt.title('Training Loss Curve')
plt.legend()
plt.show()
```

In the previous code, we demonstrated over-parameterization by introducing a multi-layer perceptron (MLP) model with a large number of hidden layers and neurons. Over-parameterization refers to the situation where the number of parameters in a model is much larger than the number of samples in the training data, which may result in the model fitting very well on the training data but performing poorly on new test data, leading to overfitting. In the code, we used the MLPClassifier with a very large architecture:

```
mlp = MLPClassifier(hidden_layer_sizes=(100, 100, 100, 100, 100), max_iter
    =1000, random_state=42)
```

- **hidden_layer_sizes=(100, 100, 100, 100, 100)**: This specifies a neural network with 5 hidden layers, each containing 100 neurons. The total number of parameters (weights and biases) can be estimated as follows:
  - For the first layer: $8 \times 100 + 100 = 900$ (8 input features plus biases).
  - For each subsequent hidden layer: $100 \times 100 + 100 = 10{,}100$.
  - For the output layer: $100 \times 1 + 1 = 101$ (binary classification with 1 neuron).
- **Sample size**: The diabetes.csv dataset typically contains around 768 samples. Thus, the model has approximately 41,401 parameters, which is significantly larger than the number of samples (768). This is a clear example of over-parameterization.

Over-parameterization can lead to the following issues:

- Overfitting: The model performs extremely well on the training data but poorly on the test data because it learns the noise and specific details of the training data rather than the underlying patterns.
- High computational cost: Training a model with too many parameters is slow and resource-intensive.
- Poor generalization: The model fails to perform well on new, unseen data.

The code illustrates the consequences of over-parameterization in the following ways: (1) Training and Validation Loss Curves

```
plt.plot(mlp.loss_curve_, label='Training Loss')
plt.xlabel('Iteration')
plt.ylabel('Loss')
plt.title('Training Loss Curve')
plt.legend()
plt.show()
```

- If the model is overfitted, the accuracy on the test set will likely be much lower than that on the training set.

To mitigate over-parameterization, consider the following strategies:

- Reduce model complexity: Decrease the number of hidden layers or neurons.
- Regularization: Apply L1 or L2 regularization to constrain the model's complexity.
- Increase data volume: Obtain more data or use data augmentation to increase the number of samples.
- Early stopping: Halt training when the validation loss stops improving.

2.9 Over-Training

Another critical aspect of neural network training is over-training. When the number of training iterations is excessive, the neural network becomes overly reliant on the training data, thereby losing its ability to generalize and accurately detect test data. Essentially it is essential to identify the optimal number of iterations and halt the training process once this point is reached. To achieve this, the training process is conducted over a specified range of iterations, and the mean squared error (MSE) curves for both training and test data are plotted. The optimal iteration count is identified when the MSE curves for training and testing data begin to diverge, and the MSE of the test data starts to rise. This divergence marks the onset of over-training.

2.10 Training Based on Full Propagation

At this point, unlike the previous technique, the weights of the neural network are not updated; however, the weights are updated only once per iteration. In other words, weight changes are saved for each dataset before being applied all at once. For this purpose, the following Python script is presented below:

Python classification script through full propagation.

```
plt.plot(mlp.loss_curve_, label='Training Loss')
plt.xlabel('Iteration')
plt.ylabel('Loss')
plt.title('Training Loss Curve')
plt.legend()
plt.showimport pandas as pd
import numpy as np
import matplotlib.pyplot as plt
import seaborn as sns
%matplotlib inline

# Load the dataset
df = pd.read_csv('diabetes.csv')
df.head(10)

df.info()

# Data visualization
```

```
plt.figure(figsize=(10, 7))
ax = sns.countplot(x='Outcome', data=df, palette='viridis')
ax.spines['top'].set_visible(False)
ax.spines['right'].set_visible(False)
plt.xlabel('Has Diabetes')
plt.ylabel('Count')

for p in ax.patches:
    ax.annotate(f'{p.get_height()}', (p.get_x() + p.get_width() / 2., p.
         get_height()),
                ha='center', va='center', fontsize=11, color='gray', xytext
                    =(0, 10),
                textcoords='offset points')
plt.show()

# Data preprocessing
df_copy = df.copy(deep=True)
df_copy[['Glucose', 'BloodPressure', 'SkinThickness', 'Insulin', 'BMI']] =
     df_copy[['Glucose', 'BloodPressure', 'SkinThickness', 'Insulin', 'BMI'
     ]].replace(0, np.NaN)
df_copy.isnull().sum()

# Plotting histogram of dataset before replacing NaN values
p = df_copy.hist(figsize=(15, 15))

# Replacing NaN values with mean or median
df_copy['Glucose'] = df_copy['Glucose'].fillna(df_copy['Glucose'].mean())
df_copy['BloodPressure'] = df_copy['BloodPressure'].fillna(df_copy['
     BloodPressure'].mean())
df_copy['SkinThickness'] = df_copy['SkinThickness'].fillna(df_copy['
     SkinThickness'].median())
df_copy['Insulin'] = df_copy['Insulin'].fillna(df_copy['Insulin'].median())
df_copy['BMI'] = df_copy['BMI'].fillna(df_copy['BMI'].median())

# Plotting histogram of dataset after replacing NaN values
p = df_copy.hist(figsize=(15, 15))

# Splitting the dataset
from sklearn.model_selection import train_test_split
from sklearn.metrics import accuracy_score, classification_report
from sklearn.preprocessing import StandardScaler

X = df_copy.drop(columns='Outcome')
y = df_copy['Outcome']
X_train, X_test, y_train, y_test = train_test_split(X, y, test_size=0.20,
     random_state=0)
print('X_train size: {}, X_test size: {}'.format(X_train.shape, X_test.shape
     ))

# Feature scaling
sc = StandardScaler()
X_train = sc.fit_transform(X_train)
X_test = sc.transform(X_test)

# Training models with full propagation
results = []

# 1. K-Nearest Neighbors (K-NN)
from sklearn.neighbors import KNeighborsClassifier
knn = KNeighborsClassifier(n_neighbors=3)
knn.fit(X_train, y_train)
y_pred = knn.predict(X_test)
accuracy_knn = accuracy_score(y_test, y_pred)
```

```
results.append(accuracy_knn)
print(classification_report(y_test, y_pred))

# 2. Naive Bayes
from sklearn.naive_bayes import GaussianNB
gnb = GaussianNB()
gnb.fit(X_train, y_train)
y_pred = gnb.predict(X_test)
accuracy_gnb = accuracy_score(y_test, y_pred)
results.append(accuracy_gnb)
print(classification_report(y_test, y_pred))

# 3. Support Vector Machine (SVM)
from sklearn.svm import SVC
svm = SVC(kernel='linear', C=1.0, random_state=42)
svm.fit(X_train, y_train)
y_pred = svm.predict(X_test)
accuracy_svm = accuracy_score(y_test, y_pred)
results.append(accuracy_svm)
print(classification_report(y_test, y_pred))

# 4. Decision Tree
from sklearn.tree import DecisionTreeClassifier
dt = DecisionTreeClassifier(random_state=42)
dt.fit(X_train, y_train)
y_pred = dt.predict(X_test)
accuracy_dt = accuracy_score(y_test, y_pred)
results.append(accuracy_dt)
print(classification_report(y_test, y_pred))

# 5. Random Forest
from sklearn.ensemble import RandomForestClassifier
rf = RandomForestClassifier(n_estimators=100, random_state=42)
rf.fit(X_train, y_train)
y_pred = rf.predict(X_test)
accuracy_rf = accuracy_score(y_test, y_pred)
results.append(accuracy_rf)
print(classification_report(y_test, y_pred))

# 6. Logistic Regression
from sklearn.linear_model import LogisticRegression
log_reg = LogisticRegression(random_state=42, max_iter=10000, solver='lbfgs') # Using L-BFGS solver for full propagation
log_reg.fit(X_train, y_train)
y_pred = log_reg.predict(X_test)
accuracy_log_reg = accuracy_score(y_test, y_pred)
results.append(accuracy_log_reg)
print(classification_report(y_test, y_pred))

# Results
res = pd.DataFrame([results], columns=['K-NN', 'Naive Bayes', 'SVM', 'Decision Tree', 'Random Forest', 'Logistic Regression'])
res.head()

# Plotting results
res = res.T
res.columns = ['Accuracy']
plt.figure(figsize=(10, 6))
bars = res['Accuracy'].plot(kind='bar', color='skyblue')
plt.title('Accuracy of Classification Models')
plt.xlabel('Model')
plt.ylabel('Accuracy')
plt.xticks(rotation=45)
```

```
plt.ylim(0., 1.0)

for bar in bars.patches:
    plt.annotate(format(bar.get_height(), '.4f'),
                 (bar.get_x() + bar.get_width() / 2,
                  bar.get_height()), ha='center', va='center',
                 size=10, xytext=(0, 8),
                 textcoords='offset points')
plt.show()
```

Full Propagation Implementation:

- For the LogisticRegression model, we used the solver='lbfgs' option. This solver is suitable for small to medium-sized datasets and inherently uses the full dataset for training rather than batch-based training. This aligns with the concept of "full propagation," where the entire dataset is used in each training iteration.
- Other models like KNN, SVM, Decision Tree, and Random Forest do not involve gradient descent and are naturally trained on the full dataset without batching.

Increased Iteration Limit:

- For LogisticRegression, we set max_iter=10000 to ensure the model has sufficient iterations to converge properly. This is important for models that rely on iterative optimization processes.

Chapter 3
Recursive Least Squares (RLS) Based Neural Network Training

Abstract This chapter introduces Recursive Least Squares (RLS) as an advanced neural network training algorithm, presenting it as a specialized form of stochastic gradient descent that utilizes the inverse input autocorrelation matrix as an adaptive learning rate. The chapter provides a comprehensive mathematical derivation of the RLS algorithm, including the key equations for weight updates, covariance matrix recursion, and the role of the forgetting factor in emphasizing recent data. It demonstrates practical implementation through Python code for classification tasks using diabetes datasets, comparing RLS performance with traditional gradient descent methods and other machine learning algorithms. The theoretical foundation extends to convolutional neural networks, showing how RLS optimization can be applied to both convolutional and fully connected layers. The chapter concludes with empirical comparisons highlighting RLS's faster convergence characteristics relative to conventional gradient descent approaches.

3.1 RLS Training Techniques

The Recursive Least Squares (RLS) algorithm is a commonly used adaptive filtering algorithm for signal prediction and filtering. The RLS algorithm is implemented using MATLAB and signal prediction is performed using this algorithm. The Recursive Least Squares (RLS) optimization algorithm can be considered as a special SGD algorithm with inverse input autocorrelation matrix as the learning rate.

RLS is a popular adaptive filtering algorithm characterized by fast convergence. The algorithm recursively determines the weights that minimize a weighted linear least squares loss function based on the input signal. The algorithm is more suitable for online learning than the linear least squares algorithm.

Let $\mathbb{X}_t = \{x_1, \cdots, x_t\}$ denote all sample inputs from the starting step to the current step, and let $\widehat{\mathbb{Y}}_t^* = \{y_1^*, \cdots, y_t^*\}$ denote the corresponding target output. On this basis, the quadratic minimization problem solved by the RLS algorithm at time t is defined as.

$$w_t = \underset{w}{argmin} \frac{1}{2} \sum_{i=1}^{t} \lambda^{t-i} (y_i^* - w^T x_i)^2, \tag{3.1}$$

C. Zhang et al., *Practical Neural Networks in Python and MATLAB*,
https://doi.org/10.1007/978-3-032-14746-2_3

where w is the weight vector and $\lambda \in (0, 1]$ is the forgetting factor, which enhances the importance of the recent data relative to the old data. Setting $\nabla_w \frac{1}{2} \sum_{i=1}^{t} \lambda^{t-i} (y_i^* - w^T x_i)^2 = 0$, then, we easily obtain

$$w = \left(\sum_{i=1}^{t} \lambda^{t-i} x_i x_i^T \right)^{-1} \sum_{i=1}^{t} \lambda^{t-i} x_1 y_i^*, \tag{3.2}$$

We define At and bt as follows:

$$A_t = \sum_{i=1}^{t} \lambda^{t-i} x_i x_i^T, \tag{3.3}$$

$$B_t = \sum_{i=1}^{t} \lambda^{t-i} x_i y_i^T. \tag{3.4}$$

The solution wt based on (3.2) and (3.1) can be derived as

$$w_t = A_t^{-1} B_t. \tag{3.5}$$

To avoid computing the inverse of At in (3.5), we define the inverse input auto-correlation matrix $P_t = A_t^{-1}$. Equations (3.3) and (3.4) show that A_t and B_t can be computed recursively as follows:

$$A_t = \lambda A_{t-1} + X_t X_t^T, \tag{3.6}$$

$$B_t = \lambda B_{t-1} + x_t y_t^*. \tag{3.7}$$

Using the Sherman-Morrison matrix inversion formula with (3.6), we get

$$P_t = \frac{1}{\lambda} P_{(t-1)} - \frac{1}{\lambda h_t} u_t (u_t)^T, \tag{3.8}$$

where u_t and h_t are defined as follows:

$$u_t = P_{t-1} x_t, \tag{3.9}$$

$$h_t = \lambda + u_t^T x_t. \tag{3.10}$$

Substituting (3.7) and (3.8) into (3.5), we get

$$w_t = w_{t-1} - \frac{1}{h_t} u_t e_t, \tag{3.11}$$

where e_t is defined as

$$e_t = w_{t-1}^T x_t - y_t^*. \tag{3.12}$$

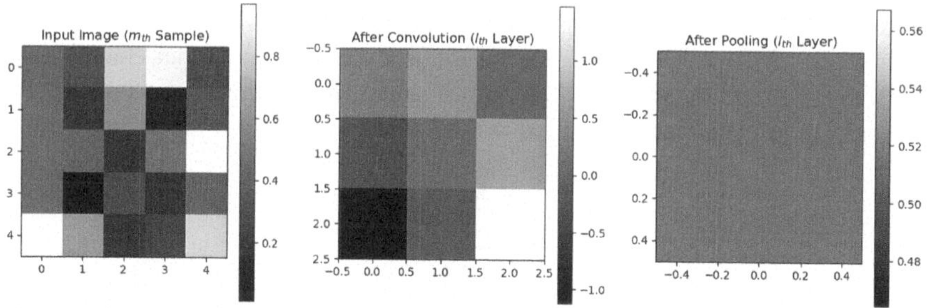

Fig. 3.1 The forward propagation learning of the m_{th} sample in the current minibatch in the l_{th} CNN layer

Finally, we obtain the RLS algorithm, which is defined by (3.8) to (3.12).

RLS optimization is a special type of SGD algorithm with an inverse input autocorrelation matrix as the learning rate. Due to the fast convergence rate of the RLS algorithm, it can be efficiently optimized for CNNs, which usually consist of an input layer followed by several convolutional layers, a pooling layer, and a fully connected layer. Since the pooling layer has no learnable weights, we only need to review the RLS optimization of the convolutional and fully connected layers. Let and denote the input and target output of the current training minibatch, respectively, and let L denote the total number of convolutional and fully connected layers. The forward propagation learning of the m_th sample in the current minibatch in the l_th CNN layer is shown in Fig. 3.1. For brevity, we omit the bias terms in each layer. Based on these notations, we briefly introduce the RLS optimization rule for CNNs. According to, the recursive update rule for the inverse input autocorrelation matrix in layer l is defined as.

$$P_t^l \approx \frac{1}{\lambda} P_{t-1}^l - \frac{k}{\lambda h_t^l} u_t^l \left(u_t^l\right)^T, \tag{3.13}$$

where $k > 0$ is the average scaling factor, and u_t^l and h_t^l are defined as follows:

$$u_t^l = P_{t-1}^l x_t^l, \tag{3.14}$$

$$h_t^l = \lambda + k\left(x_t^l\right)^T u_t^l, \tag{3.15}$$

where x_t^l s the average vector. If the l_th layer is a convolutional layer, then $x_t^l \in \mathbb{R}^{C_{l-1}H_l W_l}$ is defined as,

$$x_t^l = \frac{1}{M_t U_l V_l} \sum_{m=1}^{M_t} \sum_{u=1}^{U_l} \sum_{v=1}^{V_l} \left(flatten(R_{t(m,:,u,v,:,:)}^{l-1}\right)^T, \tag{3.16}$$

where H_l and W_l denote the height and width of the filter, M_t denotes the current batch size, and U_l and V_l denote the height and width of the output channel, and $flatten(\cdot)$ denotes the shaping of the given matrix or tensor into a column vector. If layer l is a fully connected layer, x_t^l is defined as

$$x_t^l = \frac{1}{M_t} \sum_{m=1}^{M_t} \left(Y_{t(m,:)}^{l-1} \right)^T. \tag{3.17}$$

Note that if the layer is preceded by a convolutional or pooling layer, $Y_{t(m,:)}^{l-1}$ will represent the flattened vector of the output of the previous layer. In the RLS optimization algorithm, the tensor is converted to a matrix W_{t-1}^l by $W_{t-1(:,j)}^l = flatten(W_{t-1(:,j,:,:)}^l)$ definition. Additionally, the algorithm uses momentum terms to speed up convergence. Therefore, the recursive update rule, regardless of whether the l_{th} layer is convolutional or fully connected, is defined as follows:

$$\Psi_t^l = \alpha \Psi_{t-1}^l - \frac{\eta^l}{h_t^l} P_{t-1}^l \nabla_{W_{t-1}^l}, \tag{3.18}$$

$$W_t^l \approx W_{t-1}^l + \Psi_t^l, \tag{3.19}$$

where Ψ_t^l is the velocity matrix of layer l at step t, α is the momentum factor, $\eta^l > 0$ is the gradient scaling factor, and $\nabla_{W_{t-1}^l}$ is the equivalent gradient of the linear output loss function L_t with respect to W_{t-1}^l. L_t is defined as

$$L_t = \frac{1}{2M_t} \left\| Z_t^L - Z_t^* \right\|_F^2, \tag{3.20}$$

where $Z_t^L = f_L^{-1}(Y_t^L)$ is the linear output matrix and $Z_t^* = f_L^{-1}(Y_t^*)$ is the desired linear output matrix corresponding to Z_t^L. Note that RLS optimization assumes that $f_L(\cdot)$ is monotonic in the output layer. Moreover, RLS optimization can be used for FNNs since the above equations, in addition to (3.16), can also be used for fully connected layers, and the final part of the CNN can usually be considered as an FNN.

3.2 Implementation in Python// RLS

Consider the classification example from the previous chapter. The neural network parameters are regulated through RLS in this case.

Classification example using RLS.

```
import pandas as pd
import numpy as np
import matplotlib.pyplot as plt
import seaborn as sns
from sklearn.model_selection import train_test_split
from sklearn.preprocessing import StandardScaler
from sklearn.metrics import accuracy_score, classification_report
from sklearn.linear_model import LogisticRegression
from sklearn.neighbors import KNeighborsClassifier
from sklearn.naive_bayes import GaussianNB
from sklearn.svm import SVC
from sklearn.tree import DecisionTreeClassifier
from sklearn.ensemble import RandomForestClassifier

# Load the dataset
df = pd.read_csv('diabetes.csv')

# Data preprocessing
df_copy = df.copy(deep=True)
df_copy[['Glucose','BloodPressure','SkinThickness','Insulin','BMI']] =
    df_copy[['Glucose','BloodPressure','SkinThickness','Insulin','BMI']].
    replace(0,np.NaN)
df_copy['Glucose'] = df_copy['Glucose'].fillna(df_copy['Glucose'].mean())
df_copy['BloodPressure'] = df_copy['BloodPressure'].fillna(df_copy['
    BloodPressure'].mean())
df_copy['SkinThickness'] = df_copy['SkinThickness'].fillna(df_copy['
    SkinThickness'].median())
df_copy['Insulin'] = df_copy['Insulin'].fillna(df_copy['Insulin'].median())
df_copy['BMI'] = df_copy['BMI'].fillna(df_copy['BMI'].median())

# Split the dataset
X = df_copy.drop(columns='Outcome')
y = df_copy['Outcome']
X_train, X_test, y_train, y_test = train_test_split(X, y, test_size=0.20,
    random_state=0)

# Convert y_train and y_test to NumPy arrays
y_train = y_train.to_numpy()
y_test = y_test.to_numpy()

# Feature Scaling
sc = StandardScaler()
X_train = sc.fit_transform(X_train)
X_test = sc.transform(X_test)

# Define the RLS function
def rls_update(X, y, P, theta, lambda_):
    for i in range(len(X)):
        x = X[i].reshape(-1, 1)
        y_i = y[i]
        K = (P @ x) / (lambda_ + x.T @ P @ x)
        theta += K * (y_i - x.T @ theta)
        P = (P - K @ x.T @ P) / lambda_
    return theta, P

# Initialize parameters for RLS
lambda_ = 0.98  # Forgetting factor
P = np.eye(X_train.shape[1]) / lambda_  # Initial covariance matrix
theta = np.zeros((X_train.shape[1], 1))  # Initial weights

# Update weights using RLS
```

```
theta, P = rls_update(X_train, y_train, P, theta, lambda_)

# Predict using the trained weights
y_pred = (X_test @ theta).flatten() >= 0.5

# Evaluate the model
accuracy_rls = accuracy_score(y_test, y_pred)
print(f'Accuracy using RLS: {accuracy_rls:.4f}')
print(classification_report(y_test, y_pred))

# Compare with other models
results = []

# K-NN
knn = KNeighborsClassifier(n_neighbors=3)
knn.fit(X_train, y_train)
y_pred = knn.predict(X_test)
accuracy_knn = accuracy_score(y_test, y_pred)
results.append(accuracy_knn)

# Naive Bayes
gnb = GaussianNB()
gnb.fit(X_train, y_train)
y_pred = gnb.predict(X_test)
accuracy_gnb = accuracy_score(y_test, y_pred)
results.append(accuracy_gnb)

# SVM
svm = SVC(kernel='linear', C=1.0, random_state=42)
svm.fit(X_train, y_train)
y_pred = svm.predict(X_test)
accuracy_svm = accuracy_score(y_test, y_pred)
results.append(accuracy_svm)

# Decision Tree
dt = DecisionTreeClassifier(random_state=42)
dt.fit(X_train, y_train)
y_pred = dt.predict(X_test)
accuracy_dt = accuracy_score(y_test, y_pred)
results.append(accuracy_dt)

# Random Forest
rf = RandomForestClassifier(n_estimators=100, random_state=42)
rf.fit(X_train, y_train)
y_pred = rf.predict(X_test)
accuracy_rf = accuracy_score(y_test, y_pred)
results.append(accuracy_rf)

# Logistic Regression
log_reg = LogisticRegression(random_state=42)
log_reg.fit(X_train, y_train)
y_pred = log_reg.predict(X_test)
accuracy_log_reg = accuracy_score(y_test, y_pred)
results.append(accuracy_log_reg)

# Add RLS accuracy to results
results.append(accuracy_rls)

# Create a DataFrame for results
res = pd.DataFrame([results], columns=['K-NN', 'Naive Bayes', 'SVM', '
    Decision Tree', 'Random Forest', 'Logistic Regression', 'RLS'])
res.head()
```

```
# Plot the results
res = res.T
res.columns = ['Accuracy']
plt.figure(figsize=(10, 6))
bars = res['Accuracy'].plot(kind='bar', color='skyblue')
plt.title('Accuracy of Classification Models')
plt.xlabel('Model')
plt.ylabel('Accuracy')
plt.xticks(rotation=45)
plt.ylim(0., 1.0)

for bar in bars.patches:
    plt.annotate(format(bar.get_height(), '.4f'),
                 (bar.get_x() + bar.get_width() / 2,
                  bar.get_height()), ha='center', va='center',
                 size=10, xytext=(0, 8),
                 textcoords='offset points')
plt.show()
```

3.3 Comparison with the Gradient Descent Method

The idea of gradient descent The parameters are updated by search direction and step size. Where the search direction is the negative gradient direction of the objective function at the current position. Because this direction is the fastest descent direction. The step size determines the size of the descent along this search direction. The process of iteration is like a continuous descent to finally reach the slope. The next objective function is exemplified by the objective function of linear regression:

$$h(\theta) = \sum_{j=0}^{n} \theta_j x_j, \tag{3.21}$$

$$J(\theta) = \frac{1}{2m} \sum_{i=0}^{m} \left(y^i - h_\theta\left(x^i\right)\right)^2. \tag{3.22}$$

Batch gradient descent method

$$\frac{\partial J(\theta)}{\partial \theta_j} = -\frac{1}{m} \sum_{i=1}^{m} \left(y^i - h_\theta\left(x^i\right)\right) x_j^i, \tag{3.23}$$

$$\theta_j' = \theta_j + \frac{1}{m} \sum_{i=1}^{m} \left(y^i - h_\theta\left(x^i\right)\right) x_j^i. \tag{3.24}$$

Stochastic gradient descent method

$$\frac{\partial J(\theta)}{\partial \theta_j} = -\frac{1}{m}\sum_{i=1}^{m}\left(y^i - h_\theta\left(x^i\right)\right)x_j^i, \tag{3.25}$$

$$\theta_j{}' = \theta_j + \left(y^i - h_\theta\left(x^i\right)\right)x_j^i. \tag{3.26}$$

Add comparison with the gradient descent method.

```
import pandas as pd
import numpy as np
import matplotlib.pyplot as plt
import seaborn as sns
from sklearn.model_selection import train_test_split
from sklearn.preprocessing import StandardScaler
from sklearn.metrics import accuracy_score, classification_report
from sklearn.linear_model import LogisticRegression
from sklearn.neighbors import KNeighborsClassifier
from sklearn.naive_bayes import GaussianNB
from sklearn.svm import SVC
from sklearn.tree import DecisionTreeClassifier
from sklearn.ensemble import RandomForestClassifier

# Load the dataset
df = pd.read_csv('diabetes.csv')

# Data preprocessing
df_copy = df.copy(deep=True)
df_copy[['Glucose','BloodPressure','SkinThickness','Insulin','BMI']] =
    df_copy[['Glucose','BloodPressure','SkinThickness','Insulin','BMI']].
    replace(0,np.NaN)
df_copy['Glucose'] = df_copy['Glucose'].fillna(df_copy['Glucose'].mean())
df_copy['BloodPressure'] = df_copy['BloodPressure'].fillna(df_copy['
    BloodPressure'].mean())
df_copy['SkinThickness'] = df_copy['SkinThickness'].fillna(df_copy['
    SkinThickness'].median())
df_copy['Insulin'] = df_copy['Insulin'].fillna(df_copy['Insulin'].median())
df_copy['BMI'] = df_copy['BMI'].fillna(df_copy['BMI'].median())

# Split the dataset
X = df_copy.drop(columns='Outcome')
y = df_copy['Outcome']
X_train, X_test, y_train, y_test = train_test_split(X, y, test_size=0.20,
    random_state=0)

# Convert y_train and y_test to NumPy arrays
y_train = y_train.to_numpy()
y_test = y_test.to_numpy()

# Feature Scaling
sc = StandardScaler()
X_train = sc.fit_transform(X_train)
X_test = sc.transform(X_test)

# Define the RLS function
def rls_update(X, y, P, theta, lambda_):
    for i in range(len(X)):
        x = X[i].reshape(-1, 1)
        y_i = y[i]
```

```
        K = (P @ x) / (lambda_ + x.T @ P @ x)
        theta += K * (y_i - x.T @ theta)
        P = (P - K @ x.T @ P) / lambda_
    return theta, P

# Initialize parameters for RLS
lambda_ = 0.98  # Forgetting factor
P = np.eye(X_train.shape[1]) / lambda_  # Initial covariance matrix
theta = np.zeros((X_train.shape[1], 1))  # Initial weights

# Update weights using RLS
theta, P = rls_update(X_train, y_train, P, theta, lambda_)

# Predict using the trained weights
y_pred = (X_test @ theta).flatten() >= 0.5

# Evaluate the model
accuracy_rls = accuracy_score(y_test, y_pred)
print(f'Accuracy using RLS: {accuracy_rls:.4f}')
print(classification_report(y_test, y_pred))

# Define the Gradient Descent function for Logistic Regression
def gradient_descent(X, y, theta, alpha, iterations):
    m = len(y)
    for _ in range(iterations):
        predictions = 1 / (1 + np.exp(-X @ theta))
        error = predictions - y
        gradient = (X.T @ error) / m
        theta -= alpha * gradient
    return theta

# Initialize parameters for Gradient Descent
theta_gd = np.zeros(X_train.shape[1] + 1)  # +1 for the intercept term
alpha = 0.01  # Learning rate
iterations = 1000

# Add a column of ones to X_train and X_test for the intercept term
X_train_gd = np.hstack((np.ones((X_train.shape[0], 1)), X_train))
X_test_gd = np.hstack((np.ones((X_test.shape[0], 1)), X_test))

# Train using Gradient Descent
theta_gd = gradient_descent(X_train_gd, y_train, theta_gd, alpha, iterations
    )

# Predict using the trained weights
y_pred_gd = (X_test_gd @ theta_gd) >= 0.5

# Evaluate the model
accuracy_gd = accuracy_score(y_test, y_pred_gd)
print(f'Accuracy using Gradient Descent: {accuracy_gd:.4f}')
print(classification_report(y_test, y_pred_gd))

# Compare with other models
results = []

# K-NN
knn = KNeighborsClassifier(n_neighbors=3)
knn.fit(X_train, y_train)
y_pred = knn.predict(X_test)
accuracy_knn = accuracy_score(y_test, y_pred)
results.append(accuracy_knn)

# Naive Bayes
```

```
gnb = GaussianNB()
gnb.fit(X_train, y_train)
y_pred = gnb.predict(X_test)
accuracy_gnb = accuracy_score(y_test, y_pred)
results.append(accuracy_gnb)

# SVM
svm = SVC(kernel='linear', C=1.0, random_state=42)
svm.fit(X_train, y_train)
y_pred = svm.predict(X_test)
accuracy_svm = accuracy_score(y_test, y_pred)
results.append(accuracy_svm)

# Decision Tree
dt = DecisionTreeClassifier(random_state=42)
dt.fit(X_train, y_train)
y_pred = dt.predict(X_test)
accuracy_dt = accuracy_score(y_test, y_pred)
results.append(accuracy_dt)

# Random Forest
rf = RandomForestClassifier(n_estimators=100, random_state=42)
rf.fit(X_train, y_train)
y_pred = rf.predict(X_test)
accuracy_rf = accuracy_score(y_test, y_pred)
results.append(accuracy_rf)

# Logistic Regression
log_reg = LogisticRegression(random_state=42)
log_reg.fit(X_train, y_train)
y_pred = log_reg.predict(X_test)
accuracy_log_reg = accuracy_score(y_test, y_pred)
results.append(accuracy_log_reg)

# Add RLS and Gradient Descent accuracies to results
results.append(accuracy_rls)
results.append(accuracy_gd)

# Create a DataFrame for results
res = pd.DataFrame([results], columns=['K-NN', 'Naive Bayes', 'SVM', '
    Decision Tree', 'Random Forest', 'Logistic Regression', 'RLS', '
    Gradient Descent'])
res.head()

# Plot the results
res = res.T
res.columns = ['Accuracy']
plt.figure(figsize=(10, 6))
bars = res['Accuracy'].plot(kind='bar', color='skyblue')
plt.title('Accuracy of Classification Models')
plt.xlabel('Model')
plt.ylabel('Accuracy')
plt.xticks(rotation=45)
plt.ylim(0., 1.0)

for bar in bars.patches:
    plt.annotate(format(bar.get_height(), '.4f'),
                 (bar.get_x() + bar.get_width() / 2,
                  bar.get_height()), ha='center', va='center',
                 size=10, xytext=(0, 8),
                 textcoords='offset points')
plt.show()
```

Chapter 4
Neural Networks Training Based on Second-Order Optimization Technique

Abstract This chapter explores second-order optimization techniques for neural network training, focusing on methods that leverage curvature information beyond first-order gradients. It begins with Newton's method, detailing both basic and global variants that utilize Hessian matrices for faster convergence. The Levenberg-Marquardt (LM) algorithm is extensively covered as a hybrid approach combining gradient descent and Gauss-Newton methods, particularly effective for nonlinear least squares problems. The conjugate gradient method is presented as an efficient iterative solver for symmetric positive definite systems. The chapter emphasizes practical implementation through Python code examples, including a complete LM algorithm implementation and comparative analysis using SciPy's optimization tools. Key insights highlight the computational trade-offs between providing analytical Jacobians versus numerical approximations, demonstrating how second-order methods can achieve superior convergence characteristics despite their increased computational complexity.

4.1 Introduction

The function of the optimization algorithm is to minimize (or maximize) the loss function $E(x)$ by improving the training method. Some parameters within the model are used to calculate the degree of deviation between the true and predicted values of the target value Y in the test set, and based on these parameters, the loss function $E(x)$ is formed. For example, weight (W) and bias (b) are internal parameters that are generally used to calculate the output values and play a major role in training neural network models. The internal parameters of the model play a very important role in training the model effectively and producing accurate results.

C. Zhang et al., *Practical Neural Networks in Python and MATLAB*,
https://doi.org/10.1007/978-3-032-14746-2_4

4.2 Newton's Method

Overview of Newton's method

Newton's method is one of the optimization algorithms used most often in machine learning. The basic idea of Newton's method is to use the first-order derivatives (gradient) and second-order derivatives (Hessen matrix) at the iteration point to make a quadratic approximation to the objective function, and then take the very smallest point of the quadratic model as the new iteration point, and repeat the process until the approximate very small value that satisfies the accuracy is obtained. Newton's method is quite fast and highly approximates the optimum. Newton's method is categorized into basic Newton's method and global Newton's method.

4.2.1 Principle of Basic Newton's Method

Basic Newton's method is an algorithm that uses derivatives, and the direction of its iteration at each step is along the direction of the function value decreasing at the current point.

For example, in a one-dimensional case, an optimization function $f(x)$ must be solved. The problem of finding the extreme values of a function can be transformed into finding the derivative of the function $f'(x) = 0$.

For the function $f(x)$ performing a Taylor expansion to second order yields $f(x) = f(x_k) + f'(x_k)(x - x_k) + \frac{1}{2} f''(x_k)(x - x_k)^2$ Deriving the above equation and making it 0 gives $f'(x_k) + f''(x_k)(x - x_k) = 0$, i.e., we get $x = x_k - \frac{f'(x_k)}{f''(x_k)}$. This is the updated formula for Newton's method.

4.2.2 Global Newton Method

Newton's method stands out prominently for its rapid convergence rate and local second-order convergence property. Nevertheless, a significant limitation of the basic Newton's method is that the initial point must be sufficiently close to the minima. Otherwise, there is a high probability that the algorithm will not converge. To address this issue, the global Newton method has been introduced.

It is widely recognized that Newton's method leverages the intrinsic information of the curve, which gives it a higher likelihood of converging with fewer iterations compared to gradient descent. The following Fig. 4.1 illustrates this concept by showing the process of minimizing a target equation. The yellow curve represents the iterative solution obtained using Newton's method, while the blue curve depicts the solution derived through gradient descent.

Fig. 4.1 An example of minimizing a target equation

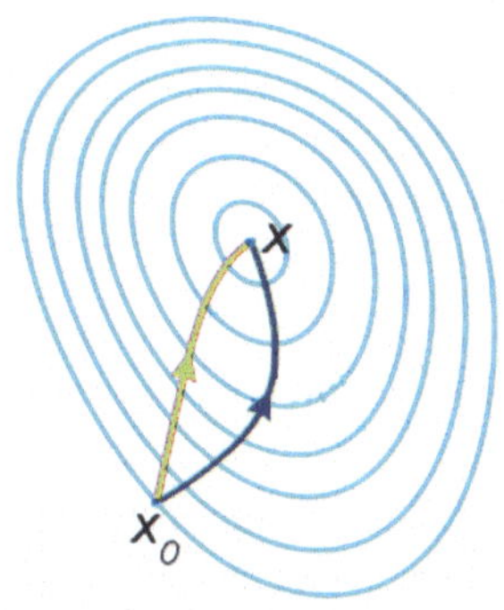

Consider the second-order Taylor expansion of the function

$$f(x + \Delta x) \approx f(x) + \nabla f^T(x)\,\Delta x + \frac{1}{2}\Delta x^T \nabla^2 f(x)\,\Delta x \tag{4.1}$$

When the above equation obtains a minimum value, there is $\Delta x = -\left[\nabla^2 f(x)\right]^{-1} \nabla f(x)$, Newton's method i.e. using the update: $\theta_{n+1} = \theta_n - \varepsilon H^{-1} \nabla f(\theta_n)$. Where θ are the network parameters, the Hessian array $H = \nabla^2 f(\theta_n)$, here we simply consider H reversible. Comparing the convergence rate of the gradient descent method on convex functions $O\left(1/\sqrt{\varepsilon}\right)$, when the initial value is close to the convergence value, Newton's method has a faster convergence rate $O(\log\log(1/\varepsilon))$.

4.3 Levenberg-Marquardt Method

The L-M method, known as the Levenberg-Marquardt method, is an estimation method for the least squares estimation of regression parameters in nonlinear regression. This method is a synthesis of the most rapid descent method and the linearization method (Taylor series). This is because the most rapid descent method is suitable for the beginning of the iteration when the parameter estimates are far from the optimum, while the linearization method, i.e., Gaussian Newton's method, is suitable for the later stages of the iteration when the parameter estimates are in the range of near-optimum values. The combination of the two methods can find the optimal value faster.

The Gauss-Newton algorithm (1809) is an old method for dealing with nonlinear least squares problems that requires the matrix columns to be full rank during the iteration process, and this condition limits its application. Levenberg (1944), but it received little attention. Later, Marquardt (1963) reintroduced it and explored it theoretically, obtaining the Levenberg-Marquardt method, or L-M method for short. Later, Fletcher (1971) improved its realization strategy and obtained the Levenberg-

Marquardt- Fletcher method. The L-M method may be more effective if it is combined with the trust domain method.

The L-M method obtains the search direction by solving the following optimization model:

$$d_k = \arg\min_{d \in R^n} \|J_k d + \gamma_k\|^2 + \mu_k \|d\|^2, \tag{4.2}$$

where $\mu_k > 0$. By the optimality condition, d_k satisfies $\left(J_k^T J_k + \mu_k I\right) d_k + J_k^T r_k = 0$, i.e. $d_k = -\left(J_k^T J_k + \mu_k I\right)^{-1} J_k^T r\left(x_k\right)$.

Using the optimality conditions of the constrained optimization problem, the L-M method can be viewed as a Gause-Newton method inspired by the trust-domain method because dk can be considered as the optimal solution of the following constrained optimization problem.

$$\min\left\{\|J_k d + \gamma_k\|^2 \,\middle|\, d \in R^n \text{ , } \|d\| \leqslant \Delta_k\right\}, \ (\Delta_k = \|d_k\|) . \tag{4.3}$$

4.3.1 *Scope of Application*

The Levenberg-Marquardt algorithm is one of the optimization algorithms. Optimization is the search for the parameter vector that minimizes the value of a function. It is used in a wide range of applications, such as: economics, management optimization, network analysis, optimal design, mechanical or electronic design, and so on.

According to the method of finding derivatives, it can be divided into 2 main categories. In the first category, if f has the form of an analytic function, it is easy to find the derivative when x is known. The second category uses numerical difference to find the derivative. Depending on the model used, they are divided into unconstrained optimization, constrained optimization, and least squares optimization.

The Levenberg-Marquardt algorithm is the most widely used nonlinear least squares algorithm. It is an algorithm that uses gradient to find the maximum (small) value, which, figuratively speaking, belongs to a kind of "mountain-climbing" method. It has the advantages of both the gradient method and the Newton method. When λ is very small, the step length is equal to the step length of Newton's method, and when λ is very large, the step length is approximately equal to the step length of the gradient descent method. The implementation of the LM algorithm is not difficult, the key is to use the model function f to make a linear approximation to the parameter vector p to be evaluated in its domain, ignoring the derivative terms above the second order, to transform it into a linear least squares problem, which has the advantages of fast convergence and so on. The LM algorithm belongs to a kind of "domain of reliance", the so-called domain of reliance method, that is to say: In optimization algorithms, the minimum value of a function is required, and the value of the objective function is required to be decreasing in each iteration, while the trust domain method, as the name suggests, starts from the initial point, assumes a maximum displacement s that can be trusted, and then solves for the true displacement by searching for the optimal point of an approximate function (quadratic) of the objective function within the region centered at the current point and radiused at s. The LM method is a kind of "trust domain method". Displacement. After the displacement is obtained, the objective function value is calculated, if it makes the decrease of the objective function value satisfy certain conditions, then it means that the displacement is reliable, then continue to calculate iteratively according to this rule; if it does not make the decrease of the objective function value satisfy certain conditions, then the range of the trust domain should be reduced and solved again.

The LM algorithm requires a partial derivation for each parameter to be estimated, so if your fitting function f is very complex or the number of parameters to be estimated is quite large, then it may not be appropriate to use the LM algorithm, and you can choose Powell's algorithm instead.

The core idea of the LM algorithm is to replace the computation of the H-matrix with the Jacobi matrix (which is easy to compute), resulting in an improvement in optimization efficiency.

LM Algorithm Flow.

```
import numpy as np

def J(x):
    return np.array([[2 * x[0], 1], [1, 2 * x[1]]])

def f(x):
    return np.array([x[0]**2 + x[1] - 1, x[0] + x[1]**2 - 2])

def F(x):
    return 0.5 * np.linalg.norm(f(x))**2

def L(x, h):
    Jx = J(x)
    A = Jx.T @ Jx
    g = Jx.T @ f(x)
    # Ensure h is a 1D array
    h = h.flatten()
    return F(x) + g.T @ h + 0.5 * h @ A @ h  # Use matrix
        multiplication with @

def LM_algorithm(x0, tau, epsilon1, epsilon2, kmax):
    k = 0  # Iteration counter
    v = 2  # Damping adjustment parameter
    x = x0  # Initial guess for the solution

    print("Initial x:", x)  # Debug: Print initial x

    # Compute initial A and g
    Jx = J(x)
    print("J(x):", Jx)  # Debug: Print J(x)

    if Jx is None:
        raise ValueError("J(x) returned None. Check the
            implementation of J(x).")

    A = Jx.T @ Jx  # A = J(x)^T * J(x)
    g = Jx.T @ f(x)  # g = J(x)^T * f(x)

```

```
    found = (np.linalg.norm(g, np.inf) <= epsilon1)  # Check if the gradient norm is below the threshold
    mu = tau * np.max(np.diag(A))  # Initial damping parameter

    while not found and k < kmax:
        k += 1

        # Solve the linear system (A + mu * I) * hlm = -g
        hlm = np.linalg.solve(A + mu * np.eye(len(x)), -g)

        # Ensure hlm is a 1D array
        hlm = hlm.flatten()

        # Check if the step size is small enough to stop
        if np.linalg.norm(hlm) <= epsilon2 * (np.linalg.norm(x) + epsilon2):
            found = True
        else:
            x_new = x + hlm  # Compute the new estimate
            e = (F(x) - F(x_new)) / (L(x, np.zeros_like(hlm)) - L(x, hlm))

            # Compute the ratio of actual to predicted reduction

            if e > 0:  # Step is acceptable
                x = x_new
                Jx = J(x)
                A = Jx.T @ Jx
                g = Jx.T @ f(x)
                found = (np.linalg.norm(g, np.inf) <= epsilon1)  # Check convergence
                mu = mu * max(1/3, 1 - (2 * e - 1)**3)  # Adjust damping parameter
                v = 2
            else:
                mu = mu * v  # Increase damping parameter
                v = 2 * v

    return x

# Example usage
x0 = np.array([1.0, 1.0])  # Initial guess
tau = 1e-3  # Initial damping factor
epsilon1 = 1e-6  # Gradient norm threshold
epsilon2 = 1e-6  # Step size threshold
kmax = 100  # Maximum number of iterations

result = LM_algorithm(x0, tau, epsilon1, epsilon2, kmax)
print("Optimized x:", result)
```

4.4 Conjugate Gradient (CG) Method

Conjugate gradient method. It is a method for solving numerical solutions of a specific set of linear equations in mathematics. Where those matrices are symmetric and positive definite. The conjugate gradient method is iterative. It is suitable for systems of linear equations with sparse matrices since these systems are too large for direct methods like Cholesky decomposition. Such systems of equations are infrequent in numerically solving partial differential equations.

The conjugate gradient method can also be used to solve unconstrained optimization problems.

In numerical linear algebra, the conjugate gradient method is an iterative method for solving a system of symmetric positive definite linear equations $Ax = b$.

The conjugate gradient method can be derived from different perspectives and contains a special case of the conjugate direction method as a method for solving optimization problems and a variant of the Arnoldi/Lanczos iteration as a method for solving eigenvalue problems.

Conjugate direction: If there are two vectors X, Y satisfying the following relation:

$X^T AY = 0$ where A is an $n \times n$ positive definite array, then X and Y are said to be conjugate concerning A. X, Y are called conjugate directions.

If $\langle X, \ Y \rangle = 0$, then X, Y are orthogonal, so it can be seen that the covariance is a generalization of orthogonality.

Conjugate vectors: Conjugate vectors play a crucial role in the conjugate gradient method. In the context of the conjugate gradient method, a set of vectors is said to be conjugate concerning a symmetric positive definite matrix A if each pair of vectors in the set satisfies the relation $X_i^T AX_j = 0$ for $i \neq j$. This means that the vectors are A-orthogonal, or conjugate, to each other. The concept of conjugate vectors is a generalization of the concept of orthogonal vectors, where the inner product is replaced by the A-inner product. The use of conjugate vectors in the conjugate gradient method allows for the efficient and effective solution of the linear system $Ax = b$, as it ensures that the search directions in the iterative process are well-chosen and lead to rapid convergence. The conjugate gradient method is particularly useful for solving large, sparse systems of linear equations where direct methods like Cholesky decomposition may be computationally expensive or impractical. By utilizing the properties of conjugate vectors and the structure of the symmetric positive definite matrix, the conjugate gradient method can provide accurate solutions with relatively low computational cost.

4.5 Implementation in Python

LM Algorithm Functions in Python

Take the following optimization function as an example: $\min \sum_{k=1}^{10} \left(2 + 2k - e^{kx_1} - e^{kx_2}\right)^2$.

Python provides an optimization function, least_squares, for solving nonlinear least squares problems. There exist two computational approaches in Python, one involving Jacobian computation and the other without it. The choice of approach impacts the efficiency and speed of convergence of the optimization process. When the Jacobian is not provided, the least_squares function must estimate it numerically.

This process is computationally intensive and can lead to a higher number of iterations, as seen in the example where 72 steps were required to reach the solution. However, when the Jacobian is analytically computed and supplied to the optimizer, the algorithm can utilize this information to navigate the solution space. This results in a significant reduction in the number of iterations, with the example demonstrating convergence in just 24 steps. The reduction in function evaluations and iterations showcases the advantages of providing an analytical Jacobian, as it allows the solver to converge more rapidly, thereby improving the overall performance of the optimization process. In conclusion, while both approaches are valid, the explicit provision of the Jacobian can greatly enhance the computational efficiency of the least_squares function.

Without Jacobian.
Step 1: Define the Objective Function.

```
import numpy as np
from scipy.optimize import least_squares

# Step1: Define the objective function
def myfun(x):
    k = np.arange(1, 11)  # Define k as 1 to 10
    F = 2 + 2 * k - np.exp(k * x[0]) - np.exp(k * x[1])  # Compute the objective function values
    return F
```

Step 2: Solve the Optimization Problem Using least_squares.

```
# Step2: Solve the optimization problem without Jacobian
x0 = [0.3, 0.4]  # Initial guess
result_without_jacobian = least_squares(myfun, x0, verbose=0)  # Invoke the optimizer

# Print the results
print("Optimal solution without Jacobian:")
print("x =", result_without_jacobian.x)
print("resnorm =", np.sum(result_without_jacobian.fun ** 2))  # Sum of squares of the residuals
```

With Jacobian.
Step 1: Define the Objective Function and Jacobian.

```
# Step1: Define the objective function and Jacobian
def myfun_with_jacobian(x):
    k = np.arange(1, 11)  # Define k as 1 to 10
    F = 2 + 2 * k - np.exp(k * x[0]) - np.exp(k * x[1])  # Compute the objective function values
    return F

def myjacobian(x):
    k = np.arange(1, 11)  # Define k as 1 to 10
    J = np.zeros((10, 2))  # Initialize the Jacobian matrix
    J[:, 0] = -k * np.exp(k * x[0])  # Partial derivatives w.r. to x[0]
    J[:, 1] = -k * np.exp(k * x[1])  # Partial derivatives w.r. to x[1]
    return J
```

Step 2: Set Up the Solver with Jacobian.

```
# Step2: Solve the optimization problem with Jacobian
x0 = [0.3, 0.4]  # Initial guess
result_with_jacobian = least_squares(myfun_with_jacobian, x0, jac=
    myjacobian, verbose=0)  # Invoke the optimizer with Jacobian

# Print the results
print("Optimal solution with Jacobian:")
print("x =", result_with_jacobian.x)
print("resnorm =", np.sum(result_with_jacobian.fun ** 2))  # Sum
    of squares of the residuals
```

Chapter 5
Neural Network Training Based on Genetic Algorithm

Abstract This chapter introduces Genetic Algorithms (GA) as a powerful evolutionary optimization technique for neural network training, particularly addressing limitations of traditional backpropagation methods like local optima convergence. It covers fundamental GA concepts including selection, crossover, and mutation operators, and their application in optimizing neural network weights and biases. The chapter demonstrates practical implementation through a comprehensive Python example solving the XOR classification problem, combining BP neural networks with genetic optimization. The implementation includes fitness evaluation, population evolution, and performance visualization. Analysis reveals steady fitness improvement over generations and discusses trade-offs between precision and recall in classification performance. The chapter concludes with practical recommendations for parameter tuning and architectural improvements to enhance GA-based neural network optimization.

5.1 Introduction

Optimization Algorithm of BP Neural Network Based on Genetic Algorithm
With the rapid development of artificial intelligence technology, neural networks, as a powerful tool, have been widely used in many fields, such as pattern recognition, predictive analysis, and decision-making. However, traditional neural network training methods, such as the BP (back propagation) algorithm, often face problems such as local optimal solutions and slow convergence speed. To solve these problems, many optimization algorithms have been proposed, among which the genetic algorithm is a very effective choice.

5.1.1 What is a Genetic Algorithm (GA)?

Genetic Algorithm (GA) is a method for solving constrained and unconstrained optimization problems based on a natural selection process that mimics biological

C. Zhang et al., *Practical Neural Networks in Python and MATLAB*,
https://doi.org/10.1007/978-3-032-14746-2_5

Table 5.1 The main differences between genetic algorithms and traditional derivative-based optimization algorithms

Classical algorithm	Genetic algorithm
Generates a single point at each iteration. The sequence of points approaches an optimal solution	Generates a population of points at each iteration. The best point in the population approaches an optimal solution
Selects the next point in the sequence by a deterministic computation	Selects the next point in the sequence by a deterministic computation

evolution. The algorithm iteratively modifies a population consisting of individual solutions. At each step, the genetic algorithm randomly selects individuals from the current population and uses them as parents to produce the next generation of offspring. After successive generations, the population "evolves" toward the optimal solution.

You can apply genetic algorithms to solve problems that don't lend themselves to standard optimization algorithms, including problems where the objective function is discontinuous, undifferentiated, stochastic, or highly nonlinear.

Genetic algorithms differ from traditional, derivative-based optimization algorithms in two main ways, as shown in the (Table 5.1).

A Genetic Algorithm is an optimization algorithm inspired by the theory of natural selection and genetics, which searches for the optimal solution of a problem by simulating the operations of selection, crossover, and mutation in biological evolution. Compared with traditional optimization algorithms, genetic algorithms have the advantages of strong global search capability and the ability to handle multi-peak functions. Combining a genetic algorithm with a BP neural network can overcome the local optimal solution problem of the BP neural network and improve the training efficiency and accuracy of the network.

The specific implementation process of the BP neural network optimization algorithm based on the genetic algorithm includes the following steps:

1. **Encoding**: convert the weights and thresholds of the neural network into binary codes to form the initial population.
2. **Decoding**: decode the binary code into actual weights and thresholds for the neural network training.
3. **Genetic operation**: Evolution of the population through selection, crossover, and mutation operations to produce a new population.
4. **Network training**: The new population is used to train the neural network and improve the network's performance.

Through experiments, it is verified that the BP neural network optimization algorithm based on a genetic algorithm has higher training efficiency and accuracy compared with the traditional BP algorithm. In terms of convergence and stability, the BP neural network optimization algorithm based on a genetic algorithm also shows obvious advantages. In addition, the algorithm has a wide range of application prospects and can be used for the training of various types of neural networks, such as deep learning networks, convolutional neural networks, and so on.

The BP neural network optimization algorithm based on a genetic algorithm has many advantages, such as improving training efficiency and accuracy and overcoming the local optimal solution problem. However, the algorithm also has some potential problems, such as the accuracy and efficiency of encoding and decoding and the setting of parameters such as population size and evolutionary generations in genetic operations. To better apply the BP neural network optimization algorithm based on the genetic algorithm, future research directions can include the following aspects:

1. **Improve the encoding and decoding methods**: The current commonly used encoding and decoding methods may have certain errors and efficiency problems, and it is of great significance to study more accurate and efficient encoding and decoding methods to improve the performance of the algorithm.
2. **Optimize genetic operation**: The setting of parameters such as population size and evolutionary generations in genetic operation may directly affect the performance of the algorithm, and it is a future research direction to study how to reasonably set these parameters to obtain better training results.
3. **Combine with other optimization algorithms**: In addition to the genetic algorithm, there are many other optimization algorithms, such as particle swarm optimization algorithm, simulated annealing algorithm, and so on. To study how to combine the BP neural network optimization algorithm based on a genetic algorithm with other optimization algorithms to further improve the training effect is another important research direction.
4. **Application Expansion**: At present, the BP neural network optimization algorithm based on a genetic algorithm is mainly used in pattern recognition, predictive analysis, and other fields. The future research direction can be expanded to other fields, such as sentiment analysis, text classification, and so on, in order to further expand the application scope of the algorithm.

In conclusion, the BP neural network optimization algorithm based on a genetic algorithm is a very effective neural network training method with wide application prospects. Future research directions can include improving encoding and decoding methods, optimizing genetic operations, combining other optimization algorithms, and expanding application areas. Through continuous research and improvement, it is believed that the BP neural network optimization algorithm based on a genetic algorithm will be applied in more fields and make greater contributions to the development of artificial intelligence technology.

5.1.2 Genetic Algorithm Operators

A genetic algorithm is an optimization algorithm based on the principle of biological evolution, and its basic idea is to simulate the natural selection and genetic mechanism in the process of biological evolution. In a genetic algorithm, individuals are regarded as solutions, and through continuous genetic operations such as selection, crossover, and mutation, better individuals are gradually evolved to find the optimal solution. The principle of the genetic algorithm mainly includes the following aspects:

1. Generation of initial population: Firstly, a certain number of initial solutions are randomly generated, and these solutions constitute the initial population. Each solution is an individual that represents a possible solution to the problem.

2. Fitness evaluation: The fitness function is used to evaluate the strengths and weaknesses of each individual. Depending on the problem, the fitness function will vary. The higher the fitness of an individual, the higher the probability that it will be passed on to the next generation.

3. Selection operation: According to the result of fitness evaluation, the selection operation selects the best individuals from the current population through certain selection mechanisms (e.g., roulette selection, tournament selection, etc.) so that they will have a chance to enter the next generation.

4. Crossover operation: Through crossover operation, some of the genes of two individuals are exchanged to produce a new individual. The crossover operation mimics the process of genetic recombination in biological evolution and can produce better offspring.

5. Mutation operation: Mutation operation simulates the process of genetic mutation in biological evolution by randomly changing some genes of an individual. Mutation operation helps to maintain the diversity of the population and avoid the algorithm falling into local optimal solutions.

6. Termination conditions: When certain termination conditions are met (such as reaching the preset maximum number of evolutionary generations, individual fitness reaching the preset threshold, etc.), the algorithm stops running and outputs the best individual in the current population as the optimal solution.

5.1.3 Application of Genetic Algorithm

Because the overall search strategy and optimization search method of genetic algorithm do not rely on gradient information or other auxiliary knowledge in the calculation, but only the objective function and the corresponding fitness function that affect the search direction, genetic algorithm provides a general framework for solving the problems of complex systems, which does not depend on the specific domain of the problem and is robust to the kinds of problems, so it is widely used in many sciences,. In the following, we will introduce some of the main application areas of genetic algorithms:

Function optimization Function optimization is a classic application area of genetic algorithms and a common algorithm for performance evaluation of genetic algorithms. Many people have constructed a wide variety of complex forms of test functions: continuous and discrete functions, convex and concave functions, low-dimensional and high-dimensional functions, single-peak and multi-peak functions, and so on. Some nonlinear, multi-model, multi-objective function optimization problems it is more difficult to solve with other optimization methods, and genetic algorithms can be convenient to get better results.

Combinatorial optimization

As the size of the problem increases, the search space of combinatorial optimization problems also increases dramatically, and sometimes, it is computationally difficult to find the optimal solution using the enumeration method. For such complex problems, it has been realized that the main focus should be on seeking satisfactory solutions, and genetic algorithms are one of the best tools for seeking such satisfactory solutions. Genetic algorithms have proved to be very effective for NP problems in combinatorial optimization. For example, genetic algorithms have been successfully applied to solve the traveler's problem, the backpack problem, the packing problem, and the graph partitioning problem.

In addition, GA has also been widely used in production scheduling problems, automatic control, robotics, image processing, artificial life, genetic coding, and machine learning.

Shop Floor Scheduling

The shop floor scheduling problem is a typical NP-hard problem. The genetic algorithm as a classical intelligent algorithm is widely used in shop floor scheduling. Many scholars are committed to solving the shop floor scheduling problem by genetic algorithm, and nowadays, very fruitful results have been achieved. From the initial traditional shop scheduling (JSP) problem to the flexible job shop scheduling problem (FJSP), genetic algorithms have excellent performance, and optimal or near-optimal solutions have been obtained in many cases.

5.2 GA in Python

The process of implementing genetic algorithm-based optimization of BP neural network prediction and classification in Python can be divided into the following steps:

1. Data Preprocessing: First, we need to preprocess the data, which includes data cleaning, feature selection, data normalization, and other steps. These steps are essential for the training of neural networks as they can enhance the generalization ability of the model.

2. Build a BP Neural Network: Next, we need to build a BP neural network using Python's relevant libraries, such as TensorFlow or PyTorch. In Python, we can utilize functions and classes provided by these libraries to create a feedforward neural network architecture and then train this network.

3. Genetic Algorithm Optimization: To optimize the parameters of the neural network, we employ the genetic algorithm. In Python, there are libraries like DEAP (Distributed Evolutionary Algorithms in Python) or other optimization libraries that can be used to implement the genetic algorithm. These libraries automatically adjust the weights and bias terms of the neural network to minimize our loss function.

4. Model Evaluation: After the model training is completed, we need to use test data to evaluate the performance of the model. We can use metrics such as accuracy, recall, F1 score, etc., to assess the model's prediction and classification ability.
5. Model Application: Finally, we can apply the trained model to real-world data for prediction and classification tasks.

Key Terms in Python Genetic Algorithm Libraries
Genetic algorithms have become a highly effective tool in solving optimization problems. They imitate the principles of natural selection and genetics, evolving over generations to find the optimal solution. Fortunately, Python offers a variety of powerful genetic algorithm libraries that enable researchers and engineers to apply this strong optimization technique. These libraries are add-ons for Python that contain functions and applications for a wide range of genetic algorithms, which can be used to tackle a large number of complex optimization problems. Whether you are a scientist, engineer, or student, these libraries can assist you whenever you need to optimize a problem.

By using Python's genetic algorithm libraries, you can conveniently define fitness functions, selection operators, crossover operators, and mutation operators to fit your specific problem. Some of these libraries also provide a graphical user interface that allows users to intuitively observe the evolution of the algorithm and the final result. When dealing with real-world problems, Python's genetic algorithm libraries are highly efficient and accurate. They are capable of handling large-scale problems and finding high-quality solutions within a short period. In addition, these libraries offer extensive documentation and examples to help users get started and solve their problems quickly.

Overall, Python's genetic algorithm libraries are powerful and easy-to-use tools that can help you solve a variety of complex optimization problems. Whether you are a beginner or a professional, these libraries provide you with great support to help you achieve better results.

5.3 Example of Implementing GA in Python

The provided code implements a genetic algorithm (GA) to optimize the weights and biases of a backpropagation (BP) neural network for solving the XOR problem, a classic non-linear classification task. The XOR problem involves classifying input pairs (x_1, x_2) into one of two classes based on the following conditions:

$$y = \{1, if\ x_1 \oplus x_2 = 1; 0, if\ x_1 \oplus x_2 = 0\}, \tag{5.1}$$

where $\oplus$ denotes the XOR operation. The neural network consists of an input layer, a hidden layer with sigmoid activation, and an output layer. The GA initializes a population of potential solutions (neural network weight configurations), evaluates their fitness based on prediction accuracy, and evolves the population over generations using selection, crossover, and mutation operations to find the optimal set of weights and biases that minimizes the prediction error. The code also includes visualization of the fitness evolution over generations and displays the prediction results and a classification report to evaluate the model's performance.

Implementing GA in Python

```
import numpy as np
import random
import matplotlib.pyplot as plt
import pandas as pd

# Sigmoid activation function and its
    derivative
def sigmoid(x):
    return 1 / (1 + np.exp(-x))

def sigmoid_derivative(x):
    return x * (1 - x)

# BP Neural Network class
class BPNeuralNetwork:
    def __init__(self, input_size,
        hidden_size, output_size):
        self.input_size = input_size
        self.hidden_size = hidden_size
        self.output_size = output_size

        # Initialize weights and biases
        self.weights_input_hidden = np.
            random.uniform(-1, 1, (
            input_size, hidden_size))
        self.bias_hidden = np.random.
            uniform(-1, 1, (1, hidden_size))

        self.weights_hidden_output = np.
            random.uniform(-1, 1, (
            hidden_size, output_size))
        self.bias_output = np.random.
            uniform(-1, 1, (1, output_size))

    def forward(self, X):

        # Forward propagation
```

```
        self.hidden_input = np.dot(X, self.weights_input_hidden) + self.bias_hidden
        self.hidden_output = sigmoid(self.hidden_input)

        self.output = np.dot(self.hidden_output, self.weights_hidden_output) + self.bias_output
        return self.output

# Genetic Algorithm class
class GeneticAlgorithm:
    def __init__(self, population_size, mutation_rate, elitism_count):
        self.population_size = population_size
        self.mutation_rate = mutation_rate
        self.elitism_count = elitism_count

    def initialize_population(self, input_size, hidden_size, output_size):
        population = []
        for _ in range(self.population_size):
            # Create a BP Neural Network instance as an individual
            individual = BPNeuralNetwork(input_size, hidden_size, output_size)
            population.append(individual)
        return population

    def fitness(self, individual, X, y):
        # Evaluate the fitness of an individual (BP Neural Network)
        predictions = individual.forward(X)
        # Calculate mean squared error
        mse = np.mean((predictions - y) ** 2)
        # Fitness is the reciprocal of the error (avoid division by zero)
        return 1.0 / (mse + 1e-6)
```

```

    def select_parents(self, population, fitness_scores):
        # Select parent individuals based on fitness
        total_fitness = np.sum(fitness_scores)
        probabilities = fitness_scores / total_fitness
        # Roulette wheel selection
        parent1 = population[np.random.choice(len(population), p=probabilities)]
        parent2 = population[np.random.choice(len(population), p=probabilities)]
        return parent1, parent2

    def crossover(self, parent1, parent2):
        # Perform crossover (weight and bias mixing)
        child = BPNeuralNetwork(parent1.input_size, parent1.hidden_size, parent1.output_size)

        # Crossover weights
        cross_point = random.randint(0, parent1.weights_input_hidden.shape[0])
        child.weights_input_hidden = np.concatenate(
            (parent1.weights_input_hidden[:cross_point], parent2.weights_input_hidden[cross_point:]))

        cross_point = random.randint(0, parent1.weights_hidden_output.shape[0])
        child.weights_hidden_output = np.concatenate(
            (parent1.weights_hidden_output[:cross_point], parent2.weights_hidden_output[cross_point:]))
```

```

        # Crossover biases
        cross_point = random.randint(0, parent1.bias_hidden.shape[1])
        child.bias_hidden = np.concatenate((parent1.bias_hidden[:, :cross_point], parent2.bias_hidden[:, cross_point:]), axis=1)

        cross_point = random.randint(0, parent1.bias_output.shape[1])
        child.bias_output = np.concatenate((parent1.bias_output[:, :cross_point], parent2.bias_output[:, cross_point:]), axis=1)

        return child

    def mutate(self, individual):
        # Perform mutation on the individual's weights and biases
        if random.random() < self.mutation_rate:
            individual.weights_input_hidden += np.random.uniform(-0.1, 0.1, individual.weights_input_hidden.shape)
            individual.weights_hidden_output += np.random.uniform(-0.1, 0.1, individual.weights_hidden_output.shape)
            individual.bias_hidden += np.random.uniform(-0.1, 0.1, individual.bias_hidden.shape)
            individual.bias_output += np.random.uniform(-0.1, 0.1, individual.bias_output.shape)

    def evolve(self, population, X, y):
        # Evaluate fitness
```

```
        fitness_scores = [self.fitness(individual, X, y) for individual in population]

        # Elitism: Keep top individuals
        elite_indices = np.argsort(fitness_scores)[-self.elitism_count:]
        new_population = [population[i] for i in elite_indices]

        # Generate new population through crossover and mutation
        while len(new_population) < self.population_size:
            parent1, parent2 = self.select_parents(population, fitness_scores)
            child = self.crossover(parent1, parent2)
            self.mutate(child)
            new_population.append(child)

        return new_population, fitness_scores

# Main function
def main():
    # Data preparation
    X = np.array([[0, 0], [0, 1], [1, 0], [1, 1]])
    y = np.array([[0], [1], [1], [0]])

    # Neural network parameters
    input_size = X.shape[1]
    hidden_size = 4
    output_size = y.shape[1]

    # Genetic algorithm parameters
    population_size = 50
    mutation_rate = 0.1
    elitism_count = 5
    generations = 100

    # Initialize genetic algorithm
```

```
    ga = GeneticAlgorithm(population_size, mutation_rate, elitism_count)
    population = ga.initialize_population(input_size, hidden_size, output_size)

    # Track fitness over generations
    fitness_history = []

    for generation in range(generations):
        print(f"Generation {generation + 1}/{generations}")
        population, fitness_scores = ga.evolve(population, X, y)
        fitness_history.append(fitness_scores)

    # Use the best individual for prediction
    best_individual = population[-1]
    predictions = best_individual.forward(X)

    # Visualization
    # Plot fitness evolution
    avg_fitness = [np.mean(scores) for scores in fitness_history]
    best_fitness = [np.max(scores) for scores in fitness_history]

    plt.figure(figsize=(10, 5))
    plt.plot(range(1, generations + 1), avg_fitness, label="Average Fitness")
    plt.plot(range(1, generations + 1), best_fitness, label="Best Fitness")
    plt.xlabel("Generation")
    plt.ylabel("Fitness")
    plt.title("Fitness Evolution Over Generations")
    plt.legend()
    plt.grid()
    plt.show()

    # Prediction results table
```

```
    results = pd.DataFrame({
        "Input": list(map(str, X)),
        "Actual Output": y.flatten(),
        "Predicted Output": predictions.flatten().round(2)
    })

    print("\nPrediction Results:")
    print(results)

    # Classification report
    from sklearn.metrics import classification_report
    predicted_labels = np.round(predictions.flatten()).astype(int)
    print("\nClassification Report:")
    print(classification_report(y, predicted_labels, zero_division=1))

if __name__ == "__main__":
    main()
```

Result Analysis

1. Fitness Evolution Over Generations
The first image shows the fitness evolution over generations. Here are the key observations:

- **Average Fitness** (blue line):
  - The average fitness starts at around 1.5 and increases steadily over the generations.
  - There is some fluctuation in the early generations, which is expected as the population is still exploring the solution space.
  - By the 100th generation, the average fitness reaches around 4.5, indicating significant improvement in the population's overall performance.

- **Best Fitness** (orange line):
  - The best fitness starts at around 3.5 and increases more smoothly compared to the average fitness.
  - It shows a consistent upward trend, reaching around 5.0 by the 100th generation.
  - This indicates that the genetic algorithm is effectively finding better solutions over time.

2. Prediction Results

The second image shows the prediction results and the classification report. Here are the key observations:

- **Prediction Results Table**:
  - The table lists the input data, actual output, and predicted output.
  - The predicted output values are close to the actual output values but not exact. For example:

    For input [0, 0], the actual output is 0, and the predicted output is 0.33.
    For input [0, 1], the actual output is 1, and the predicted output is 0.40.
    For input [1, 0], the actual output is 1, and the predicted output is 0.51.
    For input [1, 1], the actual output is 0, and the predicted output is 0.37.

- **Classification Report**:
  - The classification report provides metrics for each class (0 and 1) and overall performance.
  - **Class 0**:

    Precision: 0.67
    Recall: 1.00
    F1-score: 0.80
    Support: 2

  - **Class 1**:

    Precision: 1.00
    Recall: 0.50
    F1-score: 0.67
    Support: 2

  - **Overall Metrics**:

    Accuracy: 0.75
    Macro Average:

    · Precision: 0.83
    · Recall: 0.75
    · F1-score: 0.73

Weighted Average:

- Precision: 0.83
- Recall: 0.75
- F1-score: 0.73

3. Analysis of Prediction Results

- **Performance on Class 0**:
  - The model has a high recall (1.00) for class 0, meaning it correctly identifies all actual class 0 instances.
  - However, the precision (0.67) is lower, indicating that some predicted class 0 instances are actually class 1.
- **Performance on Class 1**:
  - The model has a high precision (1.00) for class 1, meaning that all predicted class 1 instances are correct.
  - However, the recall (0.50) is lower, indicating that the model misses some actual class 1 instances.
- **Overall Performance**:
  - The overall accuracy is 0.75, which means the model correctly predicts 75
  - The macro and weighted averages show that the model performs reasonably well but has room for improvement, especially in terms of recall for class 1.

4. Recommendations for Improvement

- **Increase Population Size**:
  - A larger population size can help the genetic algorithm explore a wider range of solutions and potentially find better individuals.
- **Adjust Mutation Rate**:
  - A slightly higher mutation rate might help introduce more diversity into the population, which could improve the exploration of the solution space.
- **Increase Number of Generations**:
  - Running the genetic algorithm for more generations might allow it to converge to a better solution.

- **Feature Engineering**:
  - If the input data is more complex, consider adding more features or transforming existing features to provide more information to the model.
- **Model Tuning**:
  - Experiment with different architectures for the neural network, such as adding more hidden layers or changing the number of neurons in the hidden layer.

5.4 Conclusion

The results show that the genetic algorithm is effectively optimizing the BP neural network, as indicated by the increasing fitness values over generations. However, there is room for improvement in the model's prediction accuracy, particularly for class 1. By adjusting the genetic algorithm parameters and potentially tuning the neural network architecture, the performance can be further enhanced.

Chapter 6
Neural Network Training Based on Particle Swarm Optimization (PSO)

Abstract This chapter explores Particle Swarm Optimization (PSO) as a bio-inspired metaheuristic for neural network training, addressing limitations of gradient-based methods like local minima convergence. It presents the fundamental PSO algorithm with velocity and position update equations, and extends the methodology through integration with gray correlation analysis for network structure optimization and Bayesian regularization for enhanced generalization. The chapter demonstrates practical implementation through Python code that minimizes the Sphere function, showing PSO's effectiveness in multidimensional optimization. Experimental results reveal PSO's superior performance over standard and adaptive BP algorithms, achieving significantly lower test error rates. The comprehensive approach combines structural optimization, regularization techniques, and swarm intelligence to create robust neural network training frameworks with improved convergence and generalization capabilities.

6.1 Introduction

In the ever-evolving landscape of artificial intelligence, neural networks have established themselves as a cornerstone technology, driving advancements in various fields, such as computer vision, natural language processing, and data analytics. These powerful computational models, inspired by the structure and function of the human brain, can learn complex patterns and relationships from data, making them highly valuable for tasks ranging from image classification and speech recognition to predictive modeling and decision-making.

However, the success of neural networks is highly dependent on their training process. Traditional training methods, like backpropagation, have been the workhorses of neural network optimization for decades. Despite their widespread use, they are not without limitations. Backpropagation and its variants often face challenges such as getting stuck in local minima, slow convergence, and the need for careful tuning of learning rates and other hyperparameters. These issues can significantly impact the performance and efficiency of neural networks, especially when dealing with large-scale, high-dimensional, and complex datasets.

C. Zhang et al., *Practical Neural Networks in Python and MATLAB*,
https://doi.org/10.1007/978-3-032-14746-2_6

In recent years, there has been a growing interest in exploring alternative optimization techniques to address these challenges and further enhance the capabilities of neural networks. One such promising approach is Particle Swarm Optimization (PSO). PSO is a population-based stochastic optimization algorithm that draws inspiration from the social behavior of organisms like birds and fish. It was first introduced by Eberhart and Kennedy in 1995 and has since gained popularity in the optimization community due to its simplicity, robustness, and ability to handle a wide range of optimization problems.

The core idea behind PSO is to simulate the social and cognitive behavior of a swarm of particles moving through a multidimensional search space. Each particle represents a potential solution to the optimization problem and adjusts its position based on its own experience and the experience of its neighbors. This collaborative search mechanism allows the swarm to explore the search space efficiently and effectively, avoiding some of the pitfalls associated with traditional gradient-based methods. The application of PSO to neural network training has opened up new possibilities for improving the training process. By treating the weights and biases of the neural network as the parameters to be optimized by the PSO algorithm, we can potentially overcome the limitations of backpropagation and achieve better generalization performance, faster convergence, and more robust solutions. This chapter will provide a comprehensive overview of the integration of PSO with neural network training, covering the fundamental principles of PSO, the adaptation of PSO for neural network optimization, and the various strategies and techniques used to enhance the performance of PSO-based neural network training.

We will begin by delving into the details of the PSO algorithm, including its mathematical formulation, the role of different parameters, and the mechanisms that drive the exploration and exploitation of the search space. Then, we will discuss how PSO can be applied to the training of different types of neural networks, such as feedforward neural networks, recurrent neural networks, and convolutional neural networks. We will also explore the various hybrid approaches that combine PSO with other optimization techniques or neural network architectures to further improve the training performance.

Furthermore, this chapter will present a series of case studies and experimental results that demonstrate the effectiveness of PSO-based neural network training in different application domains. These case studies will cover a wide range of applications, from image classification and object detection to time-series forecasting and natural language processing, providing valuable insights into the practical benefits and challenges of using PSO for neural network training.

In conclusion, the integration of PSO with neural network training offers a powerful and flexible alternative to traditional training methods. By harnessing the social and cognitive behavior of particle swarms, we can unlock new levels of performance and efficiency in neural network training, paving the way for more advanced and intelligent artificial intelligence systems. This chapter aims to provide a comprehensive and in-depth exploration of this exciting field, equipping researchers and practitioners with the knowledge and tools to effectively apply PSO-based neural network training in their work.

6.2 Algorithm Formulas

6.2.1 *Particle Swarm Algorithm*

PSO is a population-based optimization algorithm, where each particle in the population represents a feasible solution to the problem and has a fitness value associated with the objective function. The particles fly at a certain speed in the search space and adjust the speed according to their own flight experience and the current state of the optimal particles, and the individuals realize the problem of searching for the optimal solution through collaboration and competition. The speed and position model update equations of the PSO algorithm are:

$$v_{id}(t+1) = wv_{id}(t) + \varphi_1 (p_{id}(t) - x_{id}(t)) + \varphi_2 (g_{id}(t) - x_{id}(t)), \tag{6.1}$$

$$x_{id}(t+1) = x_{id}(t) + v_{id}(t+1). \tag{6.2}$$

$\varphi_1 = c_1 Rand1\,()$, $\varphi_2 = c_2 Rand2\,()$, the constants c1 and c2 indicate the degree to which a particle is influenced by social knowledge and individual awareness, and are usually set to the same value; $Rand1\,()$ and $Rand2\,()$ are random numbers in the interval (0, 1); w is the inertia weight.

6.2.2 *Gray Correlation Analysis Method*

A method of measuring the degree of correlation between factors based on the degree of similarity or dissimilarity of trends between factors. Set up a systematic sequence of factors $X_0 = (x_0(1), x_0(2), \cdots, x_0(n))$, $X_1 = (x_1(1), x_1(2), \cdots, x_1(n)), \cdots$, $X_m = (x_m(1), x_m(2), \cdots, x_m(n))$. $\gamma(x_0(k), x_i(k))$ correlation coefficient of $X_0, X_1, X_2, \cdots, X_m$.

$$\gamma(x_0(k), x_i(k)) = \frac{\min_i \min_k |x_0(k) - x_i(k)| + \zeta \max_i \max_k |x_0(k) - x_i(k)|}{|x_0(k) - x_i(k)| + \zeta \max_i \max_k |x_0(k) - x_i(k)|} \tag{6.3}$$

where $\zeta \in (0, 1)$ the gray correlation between X_i and X_0 is $\gamma(X_0, X_i) = \frac{1}{n}\sum_{k=1}^{n} \gamma(x_0(k), x_i(k))$.

6.2.3 *Regularization Method*

For the network structure is determined, how to improve the generalization performance of this problem, generally used $E = \beta \cdot E_D + \alpha \cdot E_W \cdot E_D = \frac{1}{2}\sum_{k=1}^{K}$

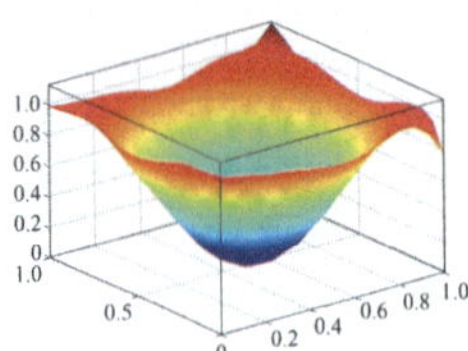

(a) LMBP prediction model.

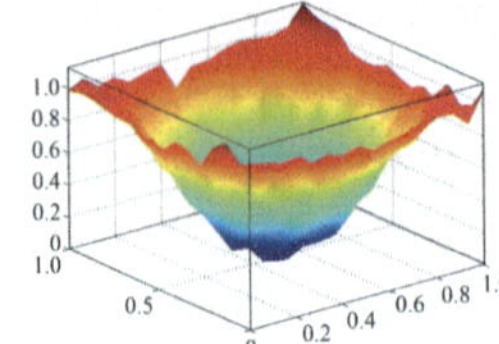

(b) LMBP prediction model after adding noise.

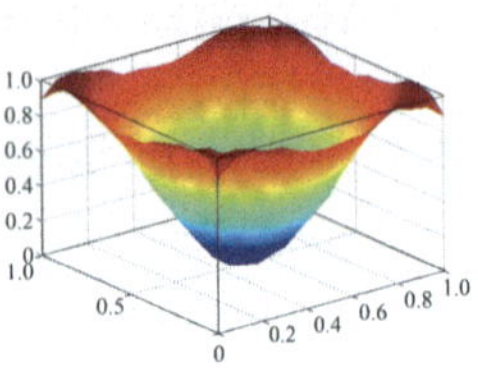

(c) Prediction model of the integrated algorithm.

Fig. 6.1 Prediction models with 256 samples

$\sum_{j=1}^{N}\left(d_{jk}-y_{jk}\right)^2$, N is the number of samples, K is the number of neural network output nodes, d_{jk} for the desired output, y_{jk} for the actual output of the network; $E_W=\frac{1}{2}\sum_{i=1}^{W}w_i^2$, E_W for the regular penalty term, w_i for the neural network's weights and thresholds, W is the number of connection weights and thresholds; α and β for the hyperparameters. The size of the neural network determines the training objective of the neural network. The shortcoming of this method is that it is difficult to determine the optimal value of the hyperparameters (Fig. 6.1).

6.2.4 Steps of Integrated Learning Algorithm for Networks Based on Particle Swarm Optimization

Optimize the network structure using PSO algorithm and gray correlation analysis:

(1) Initialize the BP network structure, transfer function, and so on.
(2) Initialize and set each parameter of the particle swarm.
(3) Calculate the artificial neural network output value, the mean square error as the fitness of each particle for search.
(4) If the maximum allowable number of iterations or the adaptation error limit is reached, the search stops; otherwise, return to (3) to continue.
(5) When the network reaches stability, use the gray correlation analysis method to calculate the correlation of the output of the hidden layer nodes of the BP neural network concerning the output of the whole network, sort them according to γ_i, and implement the deletion operation for any hidden node that satisfies $\gamma_i < \varepsilon$.
(6) If all satisfy $\gamma_i > \varepsilon$, output the number of hidden layer nodes; otherwise, turn to (2) to continue the search calculation.

Improve the generalization ability of the structure optimization network with PSO algorithm and Bayesian regularization:

(1) Initialize the number of hidden layer nodes obtained from the first part of the structure optimization to determine the BP neural network structure. Initialize the hyperparameters α, β and weights (threshold), here $\alpha=0$, $\beta=1$.

Table 6.1 Comparison of simulation results of 3 algorithms for training neural networks

Arithmetic	Number of training sessions	Minimum training error	Test error rate (%)
Standard BP algorithm	2543	0.05	11.33
Adaptive learning rate BP algorithm	96	0.05	8.67
PSO algorithm	460	0.01	4

(2) Initialize each parameter of the setup particle swarm.
(3) Calculate the artificial neural network output value as the fitness of each particle and search using the particle swarm algorithm.
(4) If the maximum allowable number of iterations or the error limit of adaptation value is reached, stop; otherwise, return to (3) to continue.
(5) Calculate $\tilde{N}\tilde{N}M$ (wMP), then calculate the number of effective parameters γ, and calculate the new estimates of hyperparameters α, β.
(6) Repeat steps (2) to (5) for different initial weight matrices until convergence.

6.3 Implementation with Python

This code implements the Particle Swarm Optimization (PSO) algorithm to solve a numerical optimization problem, specifically minimizing the Sphere function. Below is a detailed description of the problem and how the PSO algorithm tackles it:

Problem Statement

The goal is to find the minimum value of the Sphere function, a standard optimization benchmark function defined as:

$$Sphere(x) = x_1^2 + x_2^2 + \cdots + x_n^2, \tag{6.4}$$

where $x = [x_1, x_2, \ldots, x_n]$ is a vector in an n-dimensional space. The Sphere function has a global minimum at the origin ($x = [x_0, x_0, \ldots, x_0]$) with a value of 0 (Fig. 6.2).

PSO

```
import numpy as np
import matplotlib.pyplot as plt
import pandas as pd

# Objective function to minimize (Sphere
    function in this example)
```

```
def objective_function(position):
    """Calculate the value of the Sphere function for the given position."""
    return np.sum(position**2)

# PSO configuration parameters
n_particles = 30        # Number of particles in the swarm
n_dimensions = 2        # Number of dimensions to optimize
c1 = 2.0                # Cognitive (self) acceleration coefficient
c2 = 2.0                # Social (global) acceleration coefficient
w_min = 0.4             # Minimum inertia weight
w_max = 0.9             # Maximum inertia weight
max_iterations = 100    # Maximum number of iterations
lower_bound = -10.0     # Lower bound of the search space
upper_bound = 10.0      # Upper bound of the search space

# Initialize particles' positions and velocities
np.random.seed(42)  # Set random seed for reproducibility
positions = lower_bound + np.random.rand(n_particles, n_dimensions) * (upper_bound - lower_bound)
velocities = np.zeros_like(positions)

# Initialize personal best positions and values
pbest_positions = positions.copy()
pbest_values = np.array([objective_function(pos) for pos in positions])

# Initialize global best
gbest_index = np.argmin(pbest_values)
gbest_position = pbest_positions[gbest_index]
gbest_value = pbest_values[gbest_index]
```

```

# PSO main optimization loop
for iteration in range(max_iterations):
    # Update inertia weight
    w = w_max - (w_max - w_min) * (
       iteration / max_iterations)

    # Update velocities
    r1 = np.random.rand(n_particles,
       n_dimensions)
    r2 = np.random.rand(n_particles,
       n_dimensions)

    cognitive = c1 * r1 * (pbest_positions
       - positions)
    social = c2 * r2 * (gbest_position -
       positions)
    velocities = w * velocities + cognitive
        + social

    # Update positions
    positions += velocities
    # Ensure particles stay within the
       search space bounds
    positions = np.clip(positions,
       lower_bound, upper_bound)

    # Evaluate new positions
    current_values = np.array([
       objective_function(pos) for pos in
       positions])

    # Update personal bests
    better_indices = current_values <
       pbest_values
    pbest_positions[better_indices] =
       positions[better_indices].copy()
    pbest_values[better_indices] =
       current_values[better_indices]

    # Update global best
    new_gbest_index = np.argmin(
       pbest_values)
    if pbest_values[new_gbest_index] <
       gbest_value:
```

```
        gbest_value = pbest_values[new_gbest_index]
        gbest_position = pbest_positions[new_gbest_index]

    # Print progress
    print(f"Iteration {iteration+1}/{max_iterations} | Best Value: {gbest_value:.4f}")

# Visualization of particle positions
plt.figure(figsize=(8, 6))
plt.scatter(positions[:, 0], positions[:, 1], c='blue', label='Particles')
plt.scatter(gbest_position[0], gbest_position[1], c='red', marker='*', s=200, label='Global Best')
plt.title("Particle Positions in Search Space")
plt.xlabel("Dimension 1")
plt.ylabel("Dimension 2")
plt.legend()
plt.grid()
plt.show()

# Create a DataFrame to display particle information
particle_data = {
    "Particle ID": [f"Particle {i+1}" for i in range(n_particles)],
    "Position Dimension 1": positions[:, 0],
    "Position Dimension 2": positions[:, 1],
    "Objective Value": current_values,
    "Is Personal Best": [False] * n_particles
}

# Update the 'Is Personal Best' column
for i in range(n_particles):
    if np.array_equal(pbest_positions[i], positions[i]):
        particle_data["Is Personal Best"][i] = True
```

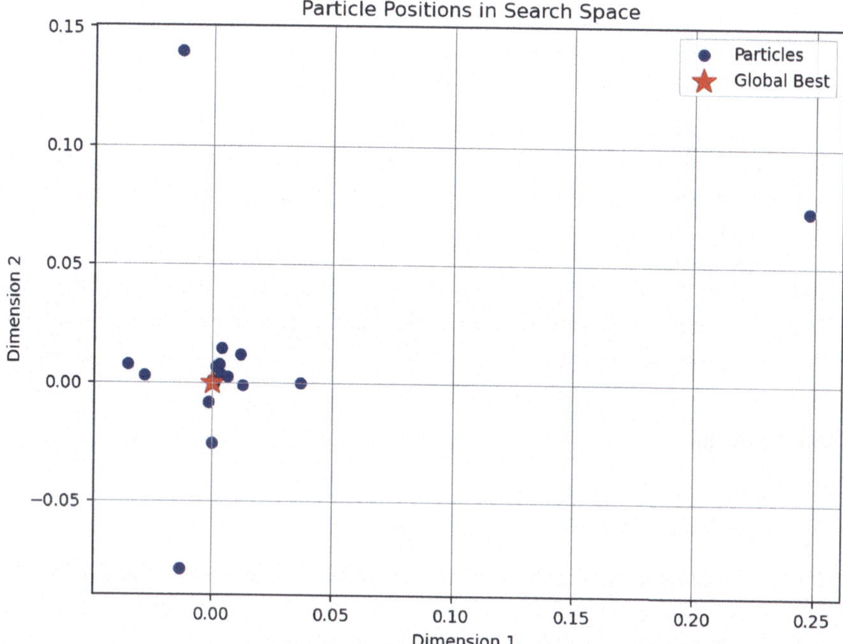

Fig. 6.2 Particle positions in search space

```

# Create and display the DataFrame
df = pd.DataFrame(particle_data)
print("\nParticle Information Table:")
print(df.to_string())

# Print final results
print("\nOptimization complete!")
print(f"Best solution found at: {
    gbest_position}")
print(f"Minimum value: {gbest_value}")
```

Result analysis:

1. Particle Positions Plot

The scatter plot shows the final positions of the particles in the search space. Key observations:

- **Particles are clustered near the origin**: This indicates that the swarm has converged toward the minimum of the Sphere function, which is at [0, 0].
- **Global Best Position**: The red star (global best) is located very close to the origin, confirming that the algorithm successfully identified the optimal solution.

2. Particle Information Table
The table provides detailed information about each particle's position, objective value, and personal best status. Key observations:

- **Objective Values**: The majority of particles have objective values close to zero, indicating they are near the global minimum.
- **Personal Bests**: Some particles (e.g., Particle 1, Particle 3, Particle 4) have achieved their personal best positions, which are marked as True in the Is Personal Best column.
- **Global Best**: The global best solution is at [2.00963099 × 10–6, 6.52989110 × 10–6], with an objective value of 4.6678094422383416 × 10–11. This is extremely close to the theoretical minimum value of 0.

3. Final Results

- **Best Solution Found**: [2.00963099 × 10–6, 6.52989110 × 10–6].
- **Minimum Value**: 4.6678094422383416 × 10–11.

These results indicate that the PSO algorithm has successfully converged to the global minimum of the Sphere function. The extremely small objective value (≈4.67 × 10–11) confirms that the solution is practically optimal.

Chapter 7
UKF-Based Neural Network Training

Abstract This chapter introduces the Unscented Kalman Filter (UKF) as an advanced neural network training methodology that combines unscented transformation with Kalman filtering principles. Unlike Extended Kalman Filters that linearize nonlinear functions, UKF employs sigma points to accurately propagate statistical distributions through nonlinear systems. The chapter presents a comprehensive mathematical formulation of UKF algorithms, including state transition models, measurement functions, and covariance updates. A practical implementation demonstrates UKF's application in target tracking scenarios with radar measurements in polar coordinates, showcasing its superiority in handling nonlinear observation models. The Python implementation using FilterPy library provides hands-on experience with UKF parameter tuning, sigma point generation, and real-time state estimation, establishing UKF as a robust alternative for neural network training in complex nonlinear environments.

7.1 UKF Algorithm

UKF (Unscented Kalman Filter), the Chinese interpretation is lossless Kalman filter, trace-free Kalman filter, or de-aromatized Kalman filter. It is a combination of lossless transformation (UT) and a standard Kalman filtering system, which makes the nonlinear system equations applicable to the standard Kalman filtering system under the assumption of linearity through lossless transformation.

Unlike the EKF (Extended Kalman Filter), the UKF makes the nonlinear system equations applicable to the standard Kalman filtering system under linear assumptions through lossless transformations, rather than realizing recursive filtering by linearizing the nonlinear functions, as in the case of the EKF. There are two theoretical foundations for target tracking, namely data correlation and Kalman filtering techniques. Since the state model and measurement model of the tracking system are mostly nonlinear in practical target tracking, the nonlinear filtering method is used.

Fuzzy neural networks are not only good at utilizing their empirical knowledge, but also introduce the learning mechanism of neural networks, which makes the fuzzy neural networks have the characteristics of strong reasoning ability, adaptive

C. Zhang et al., *Practical Neural Networks in Python and MATLAB*,
https://doi.org/10.1007/978-3-032-14746-2_7

ability, and approximation of arbitrary continuous nonlinear mapping at the same time. At present, the hotspots of fuzzy neural network research are how to generate the optimal number of fuzzy rules and the automatic generation and pruning of fuzzy rules. Literature proposed dynamic fuzzy neural networks DFNN, structure identification and parameter estimation are carried out at the same time, and the error descent rate is used as the pruning strategy to obtain a more compact structure.

7.1.1 Description of the Simulation Problem

Consider a mass M moving in a two-dimensional plane $x - y$, whose position, velocity, and acceleration at a certain moment k can be expressed by the vector $x(k)$. Assume that M is moving in a nearly uniformly accelerated straight line in the horizontal direction (x) and a nearly uniformly accelerated straight line in the vertical direction (y). The motion in both directions has additive system noise $w(k)$, then the equation of state of the motion of the mass in the Cartesian coordinate system is:

$$x(k+1) = f(x(k)) + w(k) = F_k x(k) + w(k), \tag{7.1}$$

among them,

$$F_k = \begin{bmatrix} 1\,2\,t\,0 & \frac{t^2}{2} & 0 \\ 0\,1\,0\,t & 0 & \frac{t^2}{2} \\ 0\,0\,1\,0 & t & 0 \\ 0\,0\,0\,1 & 0 & t \\ 0\,0\,0\,0 & 1 & 0 \\ 0\,0\,0\,0 & 0 & 1 \end{bmatrix}.$$

Assuming that a radar with a coordinate position of (0, 0) measures the range and angle of M. In practice, the radar has additive noise in the measurements, and then the observation equations are in the sensor's polar coordinate system.

$$z(k) = h_k(x(k)) + v(k) = \begin{bmatrix} r_k + v_r(k) \\ \varphi_k + v_p(k) \end{bmatrix} = \begin{bmatrix} \sqrt{x_k^2 + y_k^2} + v_r(k) \\ \tan^{-1}\frac{y_k}{x_k} + v_p(k) \end{bmatrix}. \tag{7.2}$$

The model's motion observation equations are clearly nonlinear in the Cartesian coordinate system. Based on the radar measurements, we use the UKF algorithm to track the target and compare the results with the EKF algorithm.

7.1.2 Problem Analysis

UKF Filter Tracking

For nonlinear systems $\begin{cases} x(k+1) = f_k x(k) + w(k) \\ z(k) = h_k(x(k)) + v(k) \end{cases}$, let $w(k)$ have covariance array Q_k and $v(k)$ have covariance array R_k. The steps of the UKF algorithm are as follows.

(1) σ points ξ^i_{k-1k-1}, based on x_{k+1k-1} and P_{k+1}, $n+1$ points σ were generated $x_{k+1k-1}, i = 0,\ 1,\ 2,\ L2n$, and the scale parameters $\alpha = 0.01,\ k = 0,\ \beta = 2$ were taken during the UT transformation.

(2) Calculate σ the point at ξ^i_{iff}, which

$$\begin{cases} \xi^i_{ik} = f_i\left(\xi^i_{i-k-1}\right),\ i = 0,\ 1,\ 2,\ L2n \\ \widehat{x}_{ik-1} = \sum_{i=0}^{2n} \omega_i^m \xi^i_{ik} \\ P_{ik-1} = \sum_{i=0}^{2n} \omega_i^f \left(\xi^i_{ik} - \widehat{x}_{ik-1}\right)\left(\omega\xi^i_{ik} - \widehat{x}_{ik-1}\right)^T + Q_{i-1} \end{cases}$$

(3) Calculate the σ-point- $\widehat{x}_{kk-1} P_{kk-1}$ through the quantile equation for x_k, i.e.

$$\begin{aligned} &= \widehat{x}_{k|k-1}; \\ &= \widehat{x}_{k|k-1} + \left(\sqrt{(n+\lambda) P_{k|k-1}}\right)_i,\ i = 1,\ 2,\ \cdots,\ n; \\ &= \widehat{x}_{k|k-1} - \left(\sqrt{(n+\lambda) P_{k|k-1}}\right)_{i-n},\ i = n+1,\ n+2,\ \cdots,\ 2n; \end{aligned}$$

(4) Compute a one-step-ahead prediction of the output, i.e.

$$\begin{aligned} &= h_k\left(\eta_k^{(i)}\right),\ i = 1,\ 2,\ \cdots,\ 2n; \\ \widehat{z}_{k|k-1} &= \sum_{i=0}^{2n} \omega_i^{(m)} \zeta_k^{(i)}; \\ &= \sum_{i=0}^{2n} \omega_i^{(c)} \left(\zeta_k^{(i)} - \widehat{z}_{k|k-1}\right)\left(\zeta_k^{(t)} - \widehat{z}_{k|k-1}\right)^T + R_k \\ &= \sum_{i=0}^{2n} \omega_i^{(c)} \left(\zeta_k^{(i)} - \widehat{x}_{k|k-1}\right)\left(\zeta_k^{(t)} - \widehat{z}_{k|k-1}\right)^T; \end{aligned}$$

(5) After obtaining new measurements, the filtering is updated.

$$\begin{cases} \widehat{x}_{k|k} = \widehat{x}_{k|k-1} + K_k\left(z_k - \widehat{z}_{k|k-1}\right); \\ K_k = P_{\widehat{x}_k \widehat{z}_k} P^{-1}_{\widehat{z}_k}; \\ P_{k|k} = P_{k|k-1} - P_{\widehat{x}_k \widehat{z}_k} P^{-1}_{\widehat{z}_k} \left(P_{\widehat{x}_k \widehat{z}_k}\right)^T. \end{cases}$$

7.2 Implementation of UKF with Python

Key Components of the Problem

1. **Nonlinear System Dynamics**:
   - **State Transition Function**: Describes how the system's state evolves. $x_k = f(x_{k-1}, u_{k-1}, \Delta t) + w_{k-1}$.
     Here, x_k is the state vector, $f(\cdot)$ is a nonlinear function, u_k is the control input, Δt is the time step, and w_k is the process noise.
   - **Measurement Function**: Relates the system's state to the measurements. $z_k = h(x_k) + v_k$.
     Here z_k is the state vector, $h(\cdot)$ is a nonlinear function, and v_k is the measurement noise.

2. **Objective**:
   - Given a sequence of noisy measurements z_k, estimate the true state x_k of the system in real-time.

If you haven't installed filterpy yet, you can do so with the following command.

bash

```
pip install filterpy
```

Here's the code:

```
from filterpy.kalman import UnscentedKalmanFilter as UKF
from filterpy.kalman import MerweScaledSigmaPoints
import numpy as np

# Define the state transition function (nonlinear)
def state_transition_function(x, dt, *args):
    """
    Nonlinear function for state transition
    :param x: state vector (n,)
    :param dt: time step (scalar)
    :return: predicted state vector (n,)
    """
    F = np.array([[1, dt],   # Example: constant velocity model
                  [0, 1]])
    return F @ x

# Define the measurement function (nonlinear)
def measurement_function(x, *args):
    """
    Nonlinear function for measurements
    :param x: state vector (n,)
    :return: measurement vector (m,)
    """
    return np.array([x[0]])  # Example: only measuring position

# Initialize sigma points for UKF
def initialize_sigma_points(n):
```

```
    """
    Initialize sigma points generator
    :param n: dimension of state vector
    :return: sigma points generator
    """
    sigma_points = MerweScaledSigmaPoints(n=n, alpha=0.1, beta=2., kappa=0)
    return sigma_points

# Main function
def main():
    # Define dimensions
    n = 2  # State vector dimension (position, velocity)
    m = 1  # Measurement vector dimension (position)

    # Initialize sigma points
    sigma_points = initialize_sigma_points(n)

    # Initialize UKF
    ukf = UKF(dim_x=n, dim_z=m, dt=0.1,  # Time step
              fx=state_transition_function,  # State transition function
              hx=measurement_function,  # Measurement function
              points=sigma_points)       # Sigma points generator

    # Initial state and covariance matrix
    ukf.x = np.array([0., 0.])  # Initial state: position=0, velocity=0
    ukf.P = np.eye(n) * 0.1      # Initial state covariance
    ukf.R = np.eye(m) * 0.1      # Measurement noise covariance
    ukf.Q = np.eye(n) * 0.01     # Process noise covariance

    # Simulate data
    true_states = np.zeros((100, n))  # True states
    measurements = np.zeros((100, m))  # Measurements
    estimated_states = np.zeros((100, n))  # Estimated states

    # Simulation loop
    for i in range(100):
        # Update true states (example: constant acceleration)
        true_states[i] = [i * 0.1, 1.0]  # True position and velocity
        measurements[i] = true_states[i][0] + np.random.randn() * 0.1  # Simulate measurements

        # UKF prediction and update
        ukf.predict()
        ukf.update(measurements[i])
        estimated_states[i] = ukf.x

    print("True state:")
    print(true_states[-1])
    print("Estimated state:")
    print(estimated_states[-1])

if __name__ == "__main__":
    main()
```

Chapter 8
Introduction of Machine Learning Libraries in Python with Illustrative Examples

Abstract This chapter provides a comprehensive exploration of major Python machine learning libraries through practical examples spanning diverse application domains. It demonstrates air quality prediction using LSTM, CNN, XGBoost, and Prophet models for CO concentration forecasting, highlighting comparative performance metrics and temporal pattern analysis. The entertainment industry analysis employs ARIMA forecasting, clustering, and recommendation systems on movie/TV show datasets to uncover production trends and audience preferences. Natural language processing capabilities are showcased using the Pattern library for tokenization, sentiment analysis, spelling correction, and text classification. Web scraping techniques with Scrapy illustrate automated data extraction from Wikipedia for knowledge discovery. Each example integrates multiple libraries including TensorFlow, scikit-learn, pandas, and matplotlib, emphasizing practical implementation strategies, data preprocessing pipelines, and visualization techniques for real-world machine learning applications.

8.1 Example 1

Air pollution is a growing concern in urban areas, directly impacting public health and the environment. Predicting air quality enables authorities to take preventative measures to safeguard communities from dangerous exposure levels. Carbon monoxide (CO), a prevalent air pollutant largely emitted from vehicles and industrial processes, poses significant health risks, making its accurate prediction crucial. Traditional air quality forecasting methods often rely on linear models that may not effectively capture complex, non-linear patterns in pollution data. With the rise of deep learning and advanced machine learning libraries, we now have the tools to leverage sophisticated models capable of handling large datasets and uncovering intricate patterns. This study explores the use of four models-Long Short-Term Memory (LSTM), Convolutional Neural Network (CNN), XGBoost, and Prophet-to forecast CO levels. We aim to identify the most effective approach for accurate air quality forecasting by evaluating each model's performance on multiple metrics.

C. Zhang et al., *Practical Neural Networks in Python and MATLAB*,
https://doi.org/10.1007/978-3-032-14746-2_8

Previous research on air quality prediction has primarily utilized statistical and machine learning methods. Traditional time series models, such as Autoregressive Integrated Moving Average (ARIMA), can model temporal data but may not handle complex nonlinear dependencies effectively. In recent years, deep learning models, particularly LSTM and CNN, have been shown to outperform traditional models by capturing temporal dependencies and local patterns. (1) LSTM networks are widely used in sequential data tasks due to their ability to retain information over long sequences, making them effective for air quality forecasting. (2) CNN models, commonly applied in image processing, have been adapted for time series data, capturing local patterns within data segments. (3) XGBoost, a popular gradient boosting method, is known for its high predictive accuracy and efficiency with tabular data and is frequently applied to time series after feature engineering. (4) Prophet, developed by Facebook, is specifically designed to handle time series data with strong seasonal and trend components, making it effective for periodic data.

The dataset used in this study is sourced from the UCI Machine Learning Repository's "Air Quality" dataset. It contains hourly measurements of various air pollutants, including CO, along with other temporal features. The code below shows how we load and preprocess the data, removing unnecessary columns and handling missing values:

Data Loading and Preprocessing

```
import pandas as pd
import numpy as np
import zipfile
import io
import requests
from sklearn.preprocessing import MinMaxScaler
from sklearn.model_selection import train_test_split
import matplotlib.pyplot as plt
from tensorflow.keras.models import Sequential
from tensorflow.keras.layers import LSTM, Dense, Dropout,
    Conv1D, MaxPooling1D, Flatten
import xgboost as xgb
from prophet import Prophet
from statsmodels.tsa.seasonal import seasonal_decompose
from sklearn.metrics import mean_squared_error,
    mean_absolute_error, mean_absolute_percentage_error
import seaborn as sns

# Load the zip file from the URL
url = "https://archive.ics.uci.edu/ml/machine-learning-
    databases/00360/AirQualityUCI.zip"
response = requests.get(url)

# Unzip the zip file
with zipfile.ZipFile(io.BytesIO(response.content)) as z:
    with z.open('AirQualityUCI.csv') as f:
        df = pd.read_csv(f, sep=';', decimal=',', header
            =0)

# Remove the last two columns with NaN values
```

```
df = df.drop(columns=['Unnamed: 15', 'Unnamed: 16'])

# Replace incorrect values (-200) with NaN
df = df.replace(-200, np.nan)

# Replace periods with colons in the 'Time' column
df['Time'] = df['Time'].str.replace('.', ':', regex=False
    )

# Convert the 'Date' and 'Time' columns to datetime
    format
df['DateTime'] = pd.to_datetime(df['Date'] + ' ' + df['
    Time'], format='%d/%m/%Y %H:%M:%S')

# Remove the original 'Date' and 'Time' columns
df = df.drop(['Date', 'Time'], axis=1)

# Set the index to the datetime column
df.set_index('DateTime', inplace=True)

# Remove all rows containing NaN values
df = df.dropna()
```

To enhance model performance, we added temporal features, such as the hour of the day, the day of the week, and an indicator for weekends.

Feature Engineering

```
# Adding new features
df['hour'] = df.index.hour
df['day_of_week'] = df.index.dayofweek
df['is_weekend'] = df['day_of_week'].apply(lambda x: 1 if x >= 5
    else 0)
```

For effective model training, we normalize the data and split it into training and test sets.

Data Scaling and Splitting

```
# Data preparation for machine learning
target_column = 'CO(GT)'  # Target variable (e.g., CO
    concentration)
y = df[target_column].values
X = df.drop(columns=[target_column])

# Data normalization
scaler_X = MinMaxScaler()
X_scaled = scaler_X.fit_transform(X)

scaler_y = MinMaxScaler()
y_scaled = scaler_y.fit_transform(y.reshape(-1, 1))

# Splitting into training and testing data
X_train, X_test, y_train, y_test = train_test_split(X_scaled,
    y_scaled, test_size=0.2, shuffle=False)
```

Model Selection
We implemented four different models to predict CO levels: 1. Long Short-Term Memory (LSTM): A type of Recurrent Neural Network (RNN) that captures long-term dependencies in sequential data. 2. Convolutional Neural Network (CNN): Adapted to detect local patterns in time series data. 3. XGBoost: A gradient boosting algorithm known for its effectiveness with structured data, particularly after feature engineering. 4. Prophet: Designed for decomposing time series into trend and seasonal components, making it well-suited for periodic data. Each model was evaluated using a test set, and their performance was measured with MSE, MAE, and MAPE.

Detailed Overview of Models and Libraries
We employ the following models and libraries to analyze and forecast carbon monoxide (CO) concentration:

I. TensorFlow/Keras-LSTM (Long Short-Term Memory)
TensorFlow is a powerful machine learning library developed by Google, and Keras is a high-level interface built into TensorFlow that simplifies building and training neural networks. In this project, we use TensorFlow and Keras to create and train recurrent neural networks (RNN) with LSTM architecture.

II. What is LSTM?
LSTM (Long Short-Term Memory) is a type of recurrent neural network (RNN) designed to handle long-term dependencies, making it ideal for time series data. Unlike traditional RNNs, LSTMs contain specialized "memory cells" that manage which information to retain or forget at each step, helping to overcome the vanishing gradient problem and enabling them to capture long-range dependencies in data.

III. Advantages of LSTM

- Ideal for sequential data, where long-term dependencies are relevant.
- Effectively captures long-term dependencies, which is essential for time series data with seasonal or trend components.

IV. Key Parameters

- nits: Number of neurons in the LSTM layer.
- return_sequences: Specifies whether the layer returns a sequence or only the last output. Useful for creating deep LSTM architectures.
- dropout: Prevents overfitting by randomly disabling neurons during training.

V. How LSTM Works
The model is trained using past time series data to predict future values. Important dependencies are preserved through memory cells, which allow the model to capture long-term dependencies inherent in time series data.

```
# Data preparation for LSTM and CNN
def create_lstm_dataset(X, y, time_steps=60):
    """
    Function to create a dataset suitable for LSTM and CNN models.
```

```
    It creates sequences of data points with a specified time step.
    """
    X_lstm, y_lstm = [], []
    for i in range(time_steps, len(X)):
        X_lstm.append(X[i - time_steps:i])  # Append the sequence of data points
        y_lstm.append(y[i])  # Append the target value for the sequence
    return np.array(X_lstm), np.array(y_lstm)

time_steps = 60  # Number of time steps for the sequence
X_train_lstm, y_train_lstm = create_lstm_dataset(X_train, y_train, time_steps)
X_test_lstm, y_test_lstm = create_lstm_dataset(X_test, y_test, time_steps)

# 1. LSTM Model
model_lstm = Sequential()
model_lstm.add(LSTM(50, return_sequences=True, input_shape=(X_train_lstm.shape[1], X_train_lstm.shape[2])))
model_lstm.add(Dropout(0.2))  # Dropout layer to prevent overfitting
model_lstm.add(LSTM(50))  # Second LSTM layer
model_lstm.add(Dropout(0.2))  # Dropout layer
model_lstm.add(Dense(1))  # Output layer with a single neuron

model_lstm.compile(optimizer='adam', loss='mean_squared_error')  # Compile the model with Adam optimizer and MSE loss
history_lstm = model_lstm.fit(X_train_lstm, y_train_lstm, epochs=20, batch_size=32, validation_data=(X_test_lstm, y_test_lstm))

# Prediction with LSTM
y_pred_lstm = model_lstm.predict(X_test_lstm)  # Make predictions on the test set
```

TensorFlow/Keras-CNN (Convolutional Neural Network)
Although primarily used for image processing, CNN can be adapted for time series to capture local patterns over time. In this context, CNN can "detect" repetitive patterns in data, such as peaks or drops in air pollution concentration at specific times, making it useful for predicting recurring patterns.

I. Why Use CNN for Time Series?
CNN effectively captures local dependencies, making it beneficial for analyzing temporal patterns. By applying convolutional layers to time series data, the model can capture patterns within defined time intervals rather than across the entire sequence.

II. Key Parameters

- filters: Number of filters (kernels) in the convolutional layer.
- kernel_size: Size of the kernel that defines how many elements each filter will consider.
- pool_size: Pooling kernel size, which reduces the output size while preserving essential features.

III. How CNN Works

In the model, convolutional filters first process the time series to extract local patterns, pooling layers then reduce the data dimensions, and the flattened data is fed into fully connected layers to make predictions.

```
# 2. CNN Model
model_cnn = Sequential()
model_cnn.add(Conv1D(filters=64, kernel_size=2, activation='relu',
      input_shape=(X_train_lstm.shape[1], X_train_lstm.shape[2])))
model_cnn.add(MaxPooling1D(pool_size=2))
model_cnn.add(Flatten())
model_cnn.add(Dense(50, activation='relu'))
model_cnn.add(Dropout(0.2))
model_cnn.add(Dense(1))

model_cnn.compile(optimizer='adam', loss='mean_squared_error')
history_cnn = model_cnn.fit(X_train_lstm, y_train_lstm, epochs=20,
      batch_size=32, validation_data=(X_test_lstm, y_test_lstm))

# Prediction with CNN
y_pred_cnn = model_cnn.predict(X_test_lstm)
```

Prophet-Time Series Forecasting

Prophet is a time series forecasting tool developed by Facebook, specifically designed to work with data that exhibits strong seasonal patterns and trends. It is particularly useful for data with daily or weekly seasonality, making it suitable for analyzing air pollution data, which often follows such periodic trends.

I. Why Use Prophet?

Prophet automatically decomposes a time series into trend and seasonal components, simplifying the forecasting process. It is especially effective for time series with distinct trends and seasonality, allowing users to model complex temporal structures effortlessly.

II. Key Parameters

- seasonality_mode: Specifies the mode of seasonality ("additive" or "multiplicative"), allowing for flexible seasonal patterns.
- changepoint_prior_scale: Controls the model's flexibility in handling trend shifts, useful for capturing abrupt changes in data.

III. How Prophet Works

Prophet automatically identifies trends and seasonal patterns, allowing users to visualize not only the forecast but also the internal structure of the time series. This makes it a valuable tool for explaining model results and understanding time series data.

```
# 3. Prophet Model
df_prophet = pd.DataFrame({
    'ds': df.index,  # 'ds' is the column name for the datetime
        index in Prophet
    'y': df[target_column]  # 'y' is the column name for the
        target variable in Prophet
})
```

```
model_prophet = Prophet()  # Initialize the Prophet model
model_prophet.fit(df_prophet)  # Fit the model to the data

# Forecasting with Prophet
future = model_prophet.make_future_dataframe(periods=365)  # Create a dataframe for future predictions (365 days ahead)
forecast_prophet = model_prophet.predict(future)  # Generate predictions for the future dataframe
```

XGBoost—Gradient Boosting for Structured Data

XGBoost is an efficient gradient boosting algorithm that works well with structured/tabular data and can be adapted for time series. In time series forecasting, XGBoost can be applied by converting data into a "features-target" format, including lagged variables as features.

I. Why Use XGBoost?

XGBoost excels in tasks with a large number of structured features, making it ideal for time series that have been converted into a feature-target format. Its performance and flexibility allow for easy hyperparameter tuning and make it scalable for large datasets.

II. Key Parameters:

- n_estimators: Number of trees in the ensemble, controlling model complexity.
- learning_rate: Learning rate that controls each tree's contribution to the final prediction.
- max_depth: Maximum depth of trees, which affects the modelâŁ™s ability to capture complex relationships.

III. How XGBoost Works

XGBoost uses an iterative approach, where each new tree attempts to correct the errors of the previous trees. In time series, the model receives information on past values, enabling it to improve forecasting accuracy through complex patterns learned in the boosted trees.

```
# 4. XGBoost Model
def create_xgb_features(y, time_steps=60):
    """
    Function to create features for XGBoost by transforming time-series data into a supervised learning format.
    """
    X, y_out = [], []
    for i in range(time_steps, len(y)):
        X.append(y[i - time_steps:i].flatten())  # Flatten the sequence to create features
        y_out.append(y[i])  # Append the target value
    return np.array(X), np.array(y_out)

# Create features for XGBoost
X_train_xgb, y_train_xgb = create_xgb_features(y_train_lstm,
    time_steps)
X_test_xgb, y_test_xgb = create_xgb_features(y_test_lstm,
    time_steps)
```

```python

# Initialize and train the XGBoost model
model_xgb = xgb.XGBRegressor(n_estimators=100)  # XGBoost
    regressor with 100 trees
model_xgb.fit(X_train_xgb, y_train_xgb)  # Fit the model to the
    training data

# Prediction with XGBoost
y_pred_xgb = model_xgb.predict(X_test_xgb)  # Make predictions on
    the test set
```

We evaluate each model's accuracy using the MSE, MAE, and MAPE metrics:

Model Evaluation

```python
# Trimming data to the same length for metric evaluation
min_len = min(len(y_test_lstm), len(y_pred_lstm), len(y_pred_cnn),
      len(y_pred_xgb))  # Find the minimum length among all
    predictions
y_test_trimmed = y_test_lstm[-min_len:]  # Trim the actual test
    data
y_pred_lstm_trimmed = y_pred_lstm[-min_len:]  # Trim the LSTM
    predictions
y_pred_cnn_trimmed = y_pred_cnn[-min_len:]  # Trim the CNN
    predictions
y_pred_xgb_trimmed = y_pred_xgb[-min_len:]  # Trim the XGBoost
    predictions

# Inverse transformation for evaluation
y_test_inv = scaler_y.inverse_transform(y_test_trimmed)  # Inverse
      transform the actual test data
y_pred_lstm_inv = scaler_y.inverse_transform(y_pred_lstm_trimmed)
      # Inverse transform the LSTM predictions
y_pred_cnn_inv = scaler_y.inverse_transform(y_pred_cnn_trimmed)  #
      Inverse transform the CNN predictions
y_pred_xgb_inv = scaler_y.inverse_transform(y_pred_xgb_trimmed.
    reshape(-1, 1))  # Inverse transform the XGBoost predictions

# Evaluation of MSE, MAE, MAPE for each model
metrics = {
    "MSE": mean_squared_error,  # Mean Squared Error
    "MAE": mean_absolute_error,  # Mean Absolute Error
    "MAPE": mean_absolute_percentage_error  # Mean Absolute
        Percentage Error
}

for metric_name, metric_func in metrics.items():
    print(f'{metric_name} LSTM: {metric_func(y_test_inv,
        y_pred_lstm_inv)}')  # Evaluate LSTM predictions
    print(f'{metric_name} CNN: {metric_func(y_test_inv,
        y_pred_cnn_inv)}')  # Evaluate CNN predictions
    print(f'{metric_name} XGBoost: {metric_func(y_test_inv,
        y_pred_xgb_inv)}')  # Evaluate XGBoost predictions
```

Analysis of Visualizations. Visualizations are essential for interpreting the model results and understanding the dynamics of air quality forecasting. Below is an analysis of the key visualizations used in this study:

A. Prophet Forecast Plot

The forecast plot generated by the Prophet model provides a visual representation of the predicted CO levels along with trend and seasonality components. Key observations from this plot include: 1. Trends and Seasonality. Prophet captures an upward trend with recurring seasonal fluctuations, which are displayed as cyclical patterns in the forecast. This is aligned with the expected variations in air pollution, where CO levels tend to fluctuate according to daily and weekly cycles influenced by traffic and industrial activities. 2. Confidence Intervals. The widening confidence intervals in the forecast plot indicate increasing uncertainty in longer-term predictions, especially as the forecast extends beyond the observed data range. This feature is typical of Prophet and other models that decompose time series data into trends and seasonality. 3. Limitations. The forecast lacks close alignment with short-term historical fluctuations, as Prophet is designed for smoother, trend-focused predictions. This means it may not capture rapid or abrupt changes in CO levels effectively, which limits its accuracy for high-frequency forecasting tasks.

B. Comparison of Model Forecasts

The multi-line forecast plot compares actual CO levels (ground truth) with the predictions from LSTM, CNN, and XGBoost, offering insights into each model's precision and responsiveness: 1. Both LSTM and XGBoost closely follow the actual CO values, particularly in capturing peaks and troughs. This indicates that these models are effective in handling the temporal dependencies present in the dataset. LSTM's ability to retain long-term dependencies helps it track general trends, while XGBoost's flexibility with structured data allows it to quickly adapt to short-term variations. 2. XGBoost appears to be highly responsive to short-term fluctuations in CO levels, which is likely due to its gradient boosting framework that adjusts for errors iteratively, improving precision for high-frequency changes. This makes it an ideal choice for tasks where capturing fine-grained changes is crucial. 3. The CNN model, while able to detect local patterns, does not align as well with actual values, particularly during rapid changes in CO concentration. This suggests that CNN may struggle with temporal dependencies in air quality data, as it is traditionally more suited for spatial pattern recognition (e.g., in image data).

C. Heatmap of Forecasting Errors by Day and Hour (LSTM)

The error heatmap for the LSTM model provides insights into periods with the highest and lowest prediction accuracy across different times of the day and days of the week: 1. The heatmap indicates that LSTM experiences higher errors during specific times, particularly around 6–8 AM and late in the evening. These peak error times may coincide with daily rush hours, where CO levels fluctuate due to increased vehicle emissions. The model's struggle during these times suggests that additional factors (e.g., traffic or weather data) may be needed to improve prediction accuracy. 2. The LSTM model demonstrates lower error rates on weekends (days 5 and 6 on the heatmap), likely due to reduced traffic and industrial activity, leading to more stable CO levels. This finding underscores the influence of human activity on CO emissions and suggests that models could benefit from accounting for weekday vs. weekend patterns. 3. By examining error patterns across different times, we identify areas where the LSTM's accuracy could be improved. Incorporating external datasets (e.g., real-time traffic data or meteorological factors) could help the model adjust its predictions during high-error periods, enhancing forecast reliability.

```
# Analysis of seasonality and trend
decomposition = seasonal_decompose(df[target_column], model='additive', period=24)  # Decompose the target variable into trend, seasonal, and residual components
decomposition.plot()  # Plot the decomposition results
plt.show()

# Visualization of Prophet forecast with uncertainty intervals
model_prophet.plot(forecast_prophet, uncertainty=True)  # Plot the forecast with confidence intervals
plt.show()

# Visualization of forecasts from all models
plt.figure(figsize=(14, 7))
plt.plot(y_test_inv, label='True Values')  # Plot true values
plt.plot(y_pred_lstm_inv, label='LSTM')  # Plot LSTM predictions
plt.plot(y_pred_cnn_inv, label='CNN')  # Plot CNN predictions
plt.plot(y_pred_xgb_inv, label='XGBoost')  # Plot XGBoost predictions
plt.legend()
plt.show()

# Heatmap of errors for LSTM
errors = np.abs(y_test_inv - y_pred_lstm_inv)  # Calculate absolute errors for LSTM predictions
results_df = pd.DataFrame({
    'hour': df.index[-len(y_test_inv):].hour,  # Extract hours from the last part of the index
    'day_of_week': df.index[-len(y_test_inv):].dayofweek,  # Extract days of the week from the last part of the index
    'error': errors.flatten()  # Flatten the errors array
})
pivot_table = results_df.pivot_table(values='error', index='day_of_week', columns='hour')  # Create a pivot table for heatmap
plt.figure(figsize=(10, 6))
sns.heatmap(pivot_table, annot=True, fmt=".2f", cmap="coolwarm")  # Plot the heatmap
plt.title("Heatmap of LSTM Forecast Errors by Weekdays and Hours")
plt.xlabel("Hour")
plt.ylabel("Day of Week")
plt.show()
```

8.2 Example 2

This code analyzes data taken from the Kaggle platform related to movies and TV series. A dataset is used that includes the following attributes: id, title, content type: movie or TV series, description, year of release, age category, duration, genres, countries of production, number of seasons for TV series, IMDb ID, rating on IMDb, number of votes on IMDb, popularity on TMDb, tmdb_score.

By analyzing these attributes, the code aims to uncover patterns and insights related to content production, audience preferences, and trends in popularity across different platforms. The inclusion of both qualitative attributes (such as genres and descriptions) and quantitative attributes (such as IMDb ratings and vote counts) allows for a multifaceted analysis, ranging from descriptive statistics to advanced machine learning clustering and forecasting. The dataset is particularly suitable for exploring the dynamics of content creation, audience reception, and the competitive landscape of the entertainment industry.

```
import pandas as pd
import numpy as np
import matplotlib.pyplot as plt
import seaborn as sns
import warnings
import pmdarima as pm
from sklearn.feature_extraction.text import TfidfVectorizer
from sklearn.cluster import KMeans

warnings.filterwarnings("ignore")

df = pd.read_csv("/file path")

print(df.columns)
print(df.head())
```

This part of the code analyzes data about the genres of movies and TV series. First, the missing values in the "genres" column are replaced with empty lines. Then, a dictionary is created that stores the number of movies and TV shows for each genre by iterating through all rows of the dataset. After that, the genres are sorted by the total amount of content. Visualization is performed using a histogram, which shows the distribution of films and TV series by genre, allowing you to visually assess the popularity of each genre for different types of content.

From the constructed histogram, it can be observed that the number of TV series in most genres is significantly less than the number of films. This trend indicates the dominance of films in the presented dataset and may indicate the specifics of the production or consumption of content for each genre.

```
df['release_year'] = pd.to_numeric(df['release_year'], errors='coerce')
df_agg = df.groupby(['release_year', 'type']).size().unstack()

plt.figure(figsize=(12, 6))
sns.set(style='darkgrid')

sns.lineplot(data=df_agg, x=df_agg.index, y='MOVIE', label='Movies')
sns.lineplot(data=df_agg, x=df_agg.index, y='SHOW', label='Shows')

plt.title('Dynamics of Adding Movies and Shows by Year')
plt.xlabel('Release Year')
plt.ylabel('Count')
plt.legend()
plt.show()
```

This piece of code analyzes time series to predict the number of movies and TV series being released in the future. For this purpose, the ARIMA model is used, which is a standard method for analyzing time series. Let's look at the whole process, step by step, as well as the mathematics that underlies it.

The general process of analysis and forecasting

The purpose of this code is to analyze the number of films and TV series produced by year and make a forecast for the future. The following steps are used for this: Data transformation and preprocessing. Data aggregation and creation of time series. Building an ARIMA model for each type of content. Forecasting based on the constructed model. Visualization of historical data and forecasts.

```
df['release_year'] = pd.to_numeric(df['release_year'],
    errors='coerce')
df = df[df['release_year'] <= 2021]

df_agg = df.groupby(['release_year', 'type']).size().
    unstack(fill_value=0)

time_series_movies = df_agg['MOVIE']
time_series_tv_shows = df_agg['SHOW']

time_series_movies = time_series_movies.dropna()
time_series_tv_shows = time_series_tv_shows.dropna()

model_movies = pm.auto_arima(time_series_movies,
                              seasonal=True,
                              m=1,
                              trace=True,
                              error_action='ignore',
                              suppress_warnings=True)
forecast_movies, conf_int_movies = model_movies.
    predict(n_periods=5, return_conf_int=True)

model_tv_shows = pm.auto_arima(time_series_tv_shows,
                                seasonal=True,
                                m=1,
                                trace=True,
                                error_action='ignore',
                                suppress_warnings=True)
forecast_tv_shows, conf_int_tv_shows = model_tv_shows.
    predict(n_periods=5, return_conf_int=True)

lower_limit = 0
forecast_movies = np.maximum(lower_limit,
    forecast_movies)
forecast_tv_shows = np.maximum(lower_limit,
    forecast_tv_shows)

plt.figure(figsize=(12, 6))
sns.set(style='darkgrid')

sns.lineplot(data=df_agg, x=df_agg.index, y='MOVIE',
    label='Movies (History)')
sns.lineplot(data=df_agg, x=df_agg.index, y='SHOW',
    label='TV Shows (History)')

```

```
plt.plot(range(df_agg.index.max() + 1, df_agg.index.
    max() + 6), forecast_movies, label='Movies (
    Forecast)', color='blue')
plt.fill_between(range(df_agg.index.max() + 1, df_agg.
    index.max() + 6), conf_int_movies[:, 0],
    conf_int_movies[:, 1], color='blue', alpha=0.1)

plt.plot(range(df_agg.index.max() + 1, df_agg.index.
    max() + 6), forecast_tv_shows, label='TV Shows (
    Forecast)', color='orange')
plt.fill_between(range(df_agg.index.max() + 1, df_agg.
    index.max() + 6), conf_int_tv_shows[:, 0],
    conf_int_tv_shows[:, 1], color='orange', alpha
    =0.1)

plt.title('Dynamics of New Movies and TV Shows
    Additions with Forecasts (Using Data up to 2021)')
plt.xlabel('Year Added')
plt.ylabel('Count')
plt.legend()
plt.show()
```

This part of the code analyzes data related to ratings and popularity of content on the IMDb and TMDB platforms, as well as visualizing the results for various genres. First of all, data is preprocessed, including filling in missing values in several columns: the average value is used for imdb_score, tmdb_score, and tmdb_popularity, and the missing values in imdb_votes are filled with zeros.

Then, the main genre for each piece of content is extracted from the genres column, which allows you to aggregate data by genre. Using the grouping function, the following statistics are calculated for each genre.

The average rating on the IMDb platform. The average rating on the TMDB platform. The total number of votes on IMDb. Overall popularity on TMDB. Next, the results are visualized using two histograms.

A histogram of the average rating by genre for IMDb and TMDB, which allows you to identify differences in ratings on the two platforms. A histogram of the total number of votes and popularity on TMDB by genre, which shows the level of audience engagement and interest in certain genres.

```
df["imdb_score"].fillna(df["imdb_score"].mean(), inplace=True)
df["tmdb_score"].fillna(df["tmdb_score"].mean(), inplace=True)
df["imdb_votes"].fillna(0, inplace=True)
df["tmdb_popularity"].fillna(df["tmdb_popularity"].mean(), inplace
    =True)

df["genres"] = df["genres"].apply(lambda x: x.strip("[]").replace(
    "'", "").split(", ")[0])

genre_stats = df.groupby("genres").agg(
    avg_imdb_score=("imdb_score", "mean"),
    avg_tmdb_score=("tmdb_score", "mean"),
    total_imdb_votes=("imdb_votes", "sum"),
    total_tmdb_votes=("tmdb_popularity", "sum")
).reset_index()
```

```

plt.figure(figsize=(14, 6))
sns.barplot(x="genres", y="avg_imdb_score", data=genre_stats,
    color='b', alpha=0.6, label='IMDb Average Score')
sns.barplot(x="genres", y="avg_tmdb_score", data=genre_stats,
    color='orange', alpha=0.6, label='TMDB Average Score')
plt.title('Average IMDb vs TMDB Scores by Genre')
plt.xlabel('Genres')
plt.ylabel('Average Score')
plt.xticks(rotation=90)
plt.legend()
plt.show()

plt.figure(figsize=(14, 6))
sns.barplot(x="genres", y="total_imdb_votes", data=genre_stats,
    color='b', alpha=0.6, label='Total IMDb Votes')
sns.barplot(x="genres", y="total_tmdb_votes", data=genre_stats,
    color='orange', alpha=0.6, label='Total TMDB Votes')
plt.title('Total IMDb vs TMDB Votes by Genre')
plt.xlabel('Genres')
plt.ylabel('Number of Votes')
plt.xticks(rotation=90)
plt.legend()
plt.show()
```

This part of the code clusters content descriptions, uses text analysis to group by topic, and then provides recommendations based on user browsing history. Let's look at each step in more detail, including an explanation of the mathematical operations and models used in the process.

The general process of work. The purpose of this code is to analyze movie and TV series descriptions, cluster them based on thematic similarity, and build recommendations for the user based on browsing history. To do this, the code performs the following steps:

Data preprocessing: Filling in missing values in descriptions. Vectorization of descriptions using TF-IDF. Clustering of data using the K-Means algorithm. Highlighting keywords for each cluster. Generating recommendations based on user history.

```
df['description'].fillna('', inplace=True)

tfidf_vectorizer = TfidfVectorizer(stop_words='english')
tfidf_matrix = tfidf_vectorizer.fit_transform(df['description'])

num_clusters = 10
kmeans = KMeans(n_clusters=num_clusters, random_state=42)
kmeans.fit(tfidf_matrix)

df['cluster'] = kmeans.labels_

print("Keywords for each cluster:")
features = tfidf_vectorizer.get_feature_names_out()
for cluster_num in range(num_clusters):
    cluster_center = kmeans.cluster_centers_[cluster_num]
    top_features_idx = cluster_center.argsort()[-5:][::-1]
    top_features = [features[idx] for idx in top_features_idx]
    print(f"Cluster {cluster_num + 1}: {', '.join(top_features)}")

user_viewing_history = df.sample(5)['id'].tolist()
```

```
user_clusters = df[df['id'].isin(user_viewing_history)]['cluster'
    ].unique()

recommendations = df[(df['cluster'].isin(user_clusters)) & (~df['
    id'].isin(user_viewing_history))]

top_recommendations = recommendations.head(5)

print("\nUser Viewing History (Watched Movies):")
for i, movie_id in enumerate(user_viewing_history, 1):
    movie = df[df['id'] == movie_id].iloc[0]
    print(f"{i}. {movie.title} (Type: {movie.type}, Release Year:
        {movie.release_year}, Cluster: {movie.cluster})")

print("\nRecommended Movies for the User:")
for i, row in enumerate(top_recommendations.itertuples(), 1):
    print(f"{i}. {row.title} (Type: {row.type}, Release Year: {row
        .release_year}, Cluster: {row.cluster})")
```

8.3 Example 3

Pattern Library

The Pattern library is a multipurpose library capable of handling the following tasks: Natural Language Processing: Performing tasks such as tokenization, stemming, POS tagging, sentiment analysis, etc.

Data Mining: It contains APIs to mine data from sites like Twitter, Facebook, Wikipedia, etc.

Machine Learning: Contains machine learning models such as SVM, KNN, and perceptron, which can be used for classification, regression, and clustering tasks.

We will explore the use of the Pattern Library for NLP by performing tasks such as tokenization, stemming, and sentiment analysis.

Installing the Library

```
!pip install pattern
```

Pattern Library Functions for NLP

In this section, we will see some of the NLP applications of the Pattern Library.

Tokenizing, POS Tagging, and Chunking

In the NLTK and spaCy libraries, we have separate functions for tokenizing, POS tagging, and finding noun phrases in text documents. On the other hand, in the Pattern library,y there is the all-in-one parse method that takes a text string as an input parameter and returns corresponding tokens in the string, along with the POS tag.

The parse method also tells us if a token is a noun phrase or verb phrase, or subject or object. You can also retrieve lemmatized tokens by setting the lemmata parameter to True. The syntax of the parse method, along with the default values for different parameters, is as follows:

```
parse(string,
tokenize=True        # Split punctuation marks from words?
tags=True,           #Parse part-of-speech tags?(NN,JJ,
chunks=True,         #Parse chunks?(NP,VP,PNP,...
relations=False,     #Parse chunk relations?(-SBJ,-OBJ,...
lemmata=False,       #Parse lemmata?(ate =>eat)
encoding='utf-8',    # Input string encoding.
tagset=None          # Penn Treebank II(default)Or UNIVERSAL
```

Let's see the parse method in action:

```
from pattern.en import parse
from pattern.en import pprint

pprint(parse('I drove my car to the hospital yesterday', relations
    =True, lemmata=True))
```

To use the parse method, you have to import the en module from the pattern library. The en module contains English language NLP functions. If you use the pprint method to print the output of the parse method on the console, you should see the output above.

In the output, you can see the tokenized words along with their POS tag, the chunk that the tokens belong to, and the role. You can also see the lemmatized form of the tokens.

If you call the split method on the object returned by the parse method, the output will be a list of sentences, where each sentence is a list of tokens and each token is a list of words, along with the tags associated with the words.

For instance, look at the following script:

```
from pattern.en import parse
from pattern.en import pprint

print(parse('I drove my car to the hospital yesterday', relations=
    True, lemmata=True).split())
```

Pluralizing and Singularizing the Tokens

The pluralize and singularize methods are used to convert singular words to plurals and vice versa, respectively.

```
from pattern.en import pluralize, singularize

print(pluralize('leaf'))
print(singularize('thieves'))
```

Converting Adjectives to Comparative and Superlative Degrees

You can retrieve comparative and superlative degrees of an adjective using comparative and superlative functions. For instance, the comparative degree of good is better, and the superlative degree of good is best. Let's see this in action:

```
from pattern.en import comparative, superlative

print(comparative('good'))
print(superlative('good'))
```

Finding N-Grams

N-Grams refer to an "n" combination of words in a sentence. For instance, for the sentence "He goes to the hospital", 2-grams would be (He goes), (goes to), and (to hospital). N-Grams can play a crucial role in text classification and language modeling.

In the Pattern library, the ngram method is used to find all the n-grams in a text string. The first parameter to the ngram method is the text string. The number of n-grams is passed to the n parameter of the method. Look at the following example:

```
from pattern.en import ngrams

print(ngrams("He goes to hospital", n=2))
```

Finding Sentiments

Sentiment refers to an opinion or feeling towards a certain thing. The Pattern library offers functionality to find sentiment from a text string.

In Pattern, the sentiment object is used to find the polarity (positivity or negativity) of a text along with its subjectivity.

Depending upon the most commonly occurring positive (good, best, excellent, etc.) and negative (bad, awful, pathetic, etc.) adjectives, a sentiment score between 1 and −1 is assigned to the text. This sentiment score is also called the polarity.

In addition to the sentiment score, subjectivity is also returned. The subjectivity value can be between 0 and 1. Subjectivity quantifies the amount of personal opinion and factual information contained in the text. The higher subjectivity means that the text contains personal opinion rather than factual information.

```
from pattern.en import sentiment

print(sentiment("This is an excellent movie to watch. I really
    love it"))
```

The sentence "This is an excellent movie to watch. I love it" has a sentiment of 0.75, which shows that it is highly positive. Similarly, the subjectivity of 0.8 refers to the fact that the sentence is a personal opinion of the user.

Checking if a Statement is a Fact

The modality function from the Pattern library can be used to find the degree of certainty in the text string. The modality function returns a value between −1 and 1. For facts, the modality function returns a value greater than 0.5.

Here is an example of it in action:

```
from pattern.en import parse, Sentence
from pattern.en import modality

text = "Paris is the capital of France"
sent = parse(text, lemmata=True)
sent = Sentence(sent)

print(modality(sent))
```

In the script above, we first import the parse method along with the Sentence class. On the second line, we import the modality function. The parse method takes text as input and returns a tokenized form of the text, which is then passed to the Sentence class constructor. The modality method takes the Sentence class object and returns the modality of the sentence.

Since the text string "Paris is the capital of France" is a fact, the output will have a value of 1.

Similarly, the value returned by the modality method is around 0.0 for a sentence that is not certain. Look at the following script:

```
text = "I think we can complete this task"
sent = parse(text, lemmata=True)
sent = Sentence(sent)

print(modality(sent))
```

Since the string in the above example is not very certain, the modality of the above string will be 0.25.

Spelling Corrections

The suggested method can be used to find if a word is spelled correctly or not. The suggested method returns 1 if a word is 100% correctly spelled. Otherwise, the suggested method returns the possible corrections for the word along with their probability of correctness. Look at the following example:

```
from pattern.en import suggest

print(suggest("Whitle"))
```

In the script above we have a word Whitle which is incorrectly spelled. In the output, you will see possible suggestions for this word. According to the suggest method, there is a 0.64 probability that the word is "While", similarly there is a probability of 0.29 that the word is "White", and so on. Now let's spell a word correctly:

```
from pattern.en import suggest
print(suggest("Fracture"))
```

Working with Numbers

The Pattern library contains functions that can be used to convert numbers in the form of text strings into their numeric counterparts and vice versa. To convert from text to numeric representation the number function is used. Similarly to convert back from numbers to their corresponding text representation the numerals function is used. Look at the following script:

```
from pattern.en import number, numerals

print(number("one hundred and twenty two"))
print(numerals(256.390, round=2))
```

In the output, you will see 122 which is the numeric representation of text "one hundred and twenty-two". Similarly, you should see "two hundred and fifty-six point thirty-nine" which is a text representation of the number 256.390.

Remember, for numerals function we have to provide the integer value that we want our number to be rounded-off to.

The quantify function is used to get a word count estimation of the items in the list, which provides a phrase for referring to the group. If a list has 3–8 similar items, the quantify function will quantify it to "several". Two items are quantified to a "couple".

```
from pattern.en import quantify

print(quantify(['apple', 'apple', 'apple', 'banana', 'banana', '
    banana', 'mango', 'mango']))
```

Similarly, the following example demonstrates the other word count estimations.

```
from pattern.en import quantify

print(quantify({'strawberry': 200, 'peach': 15}))
print(quantify('orange', amount=1200))
```

Example of using the pattern library for a simple machine learning task, such as text classification with Naive Bayes

This code uses a small sample dataset to classify text into categories (e.g., "Positive" or "Negative") based on sentiment.

```
from pattern.en import sentiment, positive, parse
from pattern.vector import Document, Model, NB

# Sample text data
data = [
    ("I love this movie! It was fantastic and inspiring.", "
        Positive"),
    ("The food was terrible, I hated the taste.", "Negative"),
    ("Absolutely wonderful experience, I would go again.", "
        Positive"),
    ("The service was awful, very disappointed.", "Negative"),
    ("Such a pleasant day, everything went well!", "Positive"),
    ("This is the worst thing I've ever seen.", "Negative")
]

# Prepare the model and classifier
model = Model()
classifier = NB(model)

# Training the classifier
for text, label in data:
    document = Document(text, type=label)
    classifier.train(document)

# Test the classifier
test_texts = [
    "I had a great time at the event!",
    "The product is horrible, not worth buying."
]

# Make predictions
for test_text in test_texts:
    prediction = classifier.classify(Document(test_text))
    print(f"Text: '{test_text}' -> Prediction: {prediction}")
```

```

# Sentiment analysis
for text in test_texts:
    score = sentiment(text)
    print(f"Sentiment analysis of '{text}': {score}")

# Additional usage example: sentiment scoring with positive()
for text in test_texts:
    is_positive = positive(text)
    print(f"'{text}' is positive: {is_positive}")
```

Explanation
1. Training Data: We define a small dataset with text labeled as “Positive” or “Negative”.
2. Model and Classifier: We create a Model to hold documents and an instance of the Naive Bayes (NB) classifier.
3. Training: Each labeled text is added as a Document and used to train the classifier.
4. Prediction: The classifier predicts the category for new texts.
5. Sentiment Analysis: We use sentiment() and positive() to analyze the sentiment of test texts. Output This code will output predictions for each test text and give sentiment scores.

This example demonstrates text classification, sentiment analysis, and the use of the Naive Bayes classifier in the pattern library for Python. This setup can be expanded with more data for practical use.

8.4 Example 4

Scrapy

```
# Removing the output for readability
from IPython.display import clear_output
clear_output()
```

Simple Request

```
# Step 1: Install necessary libraries
!pip install scrapy
clear_output()
```

```
# Step 2: Import necessary libraries
import scrapy
from scrapy.crawler import CrawlerProcess
from scrapy.utils.project import get_project_settings
from scrapy.http import Request
clear_output()
```

```
# Step 3: Create a spider to extract the desired text
class SimpleSpider(scrapy.Spider):
    name = 'wikipedia'
    start_urls = ['https://en.wikipedia.org/wiki/Wikipedia']

    def parse(self, response):
```

```
        # Extract text using a more specific XPath
        text = response.xpath('//div[@id="mw-content-text"]//p//
            text()').getall()
        # Join the extracted text and save it
        result = ' '.join(text).strip()

        # Step 4: Save the extracted text to a .txt file
        with open('wikipedia_text.txt', 'w', encoding='utf-8') as
            f:
            f.write(result)
```

```
# Step 5: Run the spider
process = CrawlerProcess(get_project_settings())
process.crawl(SimpleSpider)
process.start()

# Clear output if running in Jupyter Notebook
try:
    from IPython.display import clear_output
    clear_output()
except ImportError:
    pass
```

```
# Step 6: Read and display the contents of the saved .txt file
try:
    with open('wikipedia_text.txt', 'r', encoding='utf-8') as f:
        content = f.read()
        print(content)
except FileNotFoundError:
    print("The file 'wikipedia_text.txt' was not found. Please
        ensure the spider has run successfully.")
```

```
import os

# Restart the runtime
os._exit(0)
```

Wikipedia Term Extraction

```
!pip install scrapy
from IPython.display import clear_output
clear_output()
```

```
import scrapy
from scrapy.crawler import CrawlerProcess
import json
from urllib.parse import quote_plus

class WikipediaSpider(scrapy.Spider):
    name = 'wikipedia_spider'

    def __init__(self, key_word='data analysis', limit=10, *args,
        **kwargs):
        super(WikipediaSpider, self).__init__(*args, **kwargs)
        self.key_word = key_word
        self.limit = limit
        self.start_urls = [self.generate_url()]

    def generate_url(self):
```

```
        # Convert the key_word into the appropriate format for the search
        formatted_keyword = quote_plus(self.key_word)
        # Construct the URL with the specified limit
        return f'https://en.wikipedia.org/w/index.php?limit=
        {self.limit}&offset=0&profile=default&search=
        {formatted_keyword}&title=Special:Search&ns0=1'

    def parse(self, response):
        # Extract links and titles from the search results
        search_results = response.css('div.mw-search-result-heading a')
        for result in search_results:
            title = result.xpath('@title').get()  # Get title from attribute
            link = response.urljoin(result.xpath('@href').get())  # Get full URL
            yield response.follow(link, self.parse_link, meta={'title': title})

    def parse_link(self, response):
        title = response.meta['title']  # Get title from metadata
        paragraphs = response.xpath('//*[@id="mw-content-text"]//p')
        extracted_terms = []

        for p in paragraphs:
            anchors = p.xpath('.//a/text()').extract()  # Extract anchor text
            for anchor in anchors:
                extracted_terms.append(anchor.strip())

        # Filter out digits
        extracted_terms = [term for term in extracted_terms if not term.isdigit()]

        # Create a dictionary with title and terms
        data = {
            'Title': title,
            'Terms': extracted_terms
        }

        # Save to JSON file
        with open('extracted_data.json', 'a') as f:
            json.dump(data, f, ensure_ascii=False)
            f.write('\n')  # Write a new line for each entry

# Run the spider with customizable parameters if __name__ ==
"__main__":
    process = CrawlerProcess()
    process.crawl(WikipediaSpider, key_word="cyber security", limit=300)

    # You can change key_word and limit here
    process.start()
```

```python
import json

with open('extracted_data.json', 'r') as f:
    for line in f:
        data = json.loads(line.strip())
        title = data['Title']
        terms = ', '.join(data['Terms'])
        print(f"{title}:\n{terms}")
        print()
```

Chapter 9
A Summary of Different Type of Neural Networks in Matlab and Python

Abstract This comprehensive chapter provides a systematic overview of 27 major neural network architectures, presenting comparative implementations in both MATLAB and Python for each type. Beginning with fundamental models like Perceptrons and Feed Forward Networks, it progresses through advanced architectures including Recurrent Neural Networks (RNN, LSTM, GRU), Autoencoders (AE, DAE, VAE), and specialized networks like Deep Residual Networks and Generative Adversarial Networks. The chapter covers probabilistic models (Boltzmann Machines, Markov Chains), self-organizing networks (Kohonen, Hopfield), and modern architectures such as Neural Turing Machines. Each network type is accompanied by practical code examples demonstrating implementation specifics, parameter configurations, and application scenarios. The dual-language approach facilitates understanding of both MATLAB's specialized neural network toolbox and Python's deep learning ecosystems (TensorFlow, PyTorch), making this an essential reference for practitioners working across different computational environments in neural network development and deployment.

Types of artificial neural networks

There are currently many types of neural networks that exist or may be in development. Neural networks can be classified depending on their structure, data flow, neurons used and their density, layers and activation filters, etc. In this section, 27 popular types of neural networks have been prepared, each of which will be briefly explained.

9.1 Perceptron Neural Network is One of the Types of Neural Networks

A perceptron neural network is one of the most basic and simplest types of neural networks used to solve classification problems. It consists of one or more perceptron neurons that are responsible for making decisions based on inputs.

C. Zhang et al., *Practical Neural Networks in Python and MATLAB*,
https://doi.org/10.1007/978-3-032-14746-2_9

Each perceptron neuron is actually a decision-making unit that consists of two main components, weights and an activation function. The weights are responsible for influencing the weight given to the inputs, and the activation function, based on the values of the weights and inputs, announces the final decision.

The architecture of a perceptron neural network can simply consist of one or more perceptron neurons. These neurons can be single-layered or multilayered. Using a training set and adjusting weights, a perceptron neural network can automatically identify and classify patterns in data.

A perceptron neural network can be useful in simpler problems where the data can be separated linearly. For example, in the problem of distinguishing between two sets of two-dimensional data that can be separated by a line, a perceptron neural network is able to correctly distinguish and classify.

Typically, perceptrons use exponential or sigmoid activation functions whose output range is limited to a specific interval. Methods such as the Gradient Descent algorithm are used to train the network parameters. Given the simplicity and computational power of the perceptron, it is usually used for simple and linear problems and is considered a machine learning algorithm. However, for more complex and non-linear problems, more advanced neural network architectures such as Deep Neural Networks are used.

The neuron processes the sum of the input values by their weights. The sum is then sent to the activation function to produce a binary output. We can see a simple example of using a perceptron neural network in MATLAB to classify data here. Note that this is just a simple example and for more complex problems you need to define the appropriate architecture and parameters.

MATLAB

```
% Definition of training data
X = [0 0; 0 1; 1 0; 1 1];    % Inputs
Y = [0; 0; 0; 1];             % Outputs

% Definition of Perceptron Neural Network
net = perceptron;
net = train(net, X', Y');

% Network test with new data
input = [0.5 0.5];

output = net(input');

% Display output
disp(output);
```

Python

```
import numpy as np

# Definition of training data
X = np.array([[0, 0], [0, 1], [1, 0], [1, 1]])  # Inputs
Y = np.array([0, 0, 0, 1])                      # Outputs

# Initialize weights and bias for the perceptron
weights = np.zeros(X.shape[1])
bias = 0

# Define the activation function (Heaviside step function)
def activation_function(x):
    return 1 if x > 0 else 0

# Define the learning rate
learning_rate = 1

# Train the perceptron
for _ in range(1000):  # Maximum number of iterations
    error = False
    for i in range(len(X)):
        # Compute the weighted sum
        weighted_sum = np.dot(X[i], weights) + bias
        # Get the prediction
        prediction = activation_function(weighted_sum)
        # Compute the error
        delta = Y[i] - prediction
        if delta != 0:
            # Update weights and bias
            weights += learning_rate * delta * X[i]
            bias += learning_rate * delta
            error = True
    if not error:
        break  # Stop training if all samples are classified correctly

# Network test with new data
new_input = np.array([0.5, 0.5])
weighted_sum = np.dot(new_input, weights) + bias
output = activation_function(weighted_sum)

# Display output
print(output)
```

In this example, we first define the training data (inputs and outputs), then create a perceptron neural network using the function. Then train, we train the network with the training data using the function. Finally, we test a new example using the trained network and get the output. In this example, the network output is shown if the input is [0.5 0.5].

The prepared file below can be a comprehensive and complete resource if you give a class presentation on various artificial neural networks.

9.2 Feed Forward Neural Network

A feed-forward neural network is a type of neural network in which information flows in one direction (from the input layer to the output layer), without any feedback. This type of neural network is used in many machine learning applications and tasks.

A feedforward neural network is a type of neural network that consists of a number of hidden layers and an output layer. Each hidden layer contains a number of neurons that process the input features. Information flows from the input layer to the hidden layers, and in each hidden layer, an operation is performed to obtain a representation of the input using weights and activation functions. Finally, the final output is extracted from the output layer.

Feed-forward neural Network is used as a powerful machine learning model. By adjusting the weights and parameters of the network, it can identify and predict complex patterns in data. This type of neural network is used in the fields of pattern recognition, image processing, machine translation, content prediction, and many other tasks.

To use a feedforward neural network, one must first define the network architecture and then train it with training data so that it can perform well on the desired tasks. Also, paying attention to parameter tuning and optimization algorithms is essential to achieve better network performance.

MATLAB

```
% Definition of training data
X = [ 0 0; 0 1; 1 0; 1 1 ] ;    % Inputs
Y = [ 0; 1; 1; 0 ] ;             % Outputs
% Network architecture definition
net = feedforwardnet ( 3 ) ;     % Feedforward neural network with one hidden
      layer consisting of 10 neurons
% Network training
net = train (net, X', Y') ;
% Network test with new data
input = [ 0.5 0.5 ] ;
output = net ( input') ;
% Display output
disp ( output ) ;
```

Python

```
import numpy as np
import tensorflow as tf
from tensorflow import keras

# Definition of training data
X = np.array([[0, 0], [0, 1], [1, 0], [1, 1]], dtype=np.float32)  # Inputs
Y = np.array([[0], [1], [1], [0]], dtype=np.float32)              # Outputs

# Define the model architecture
model = keras.Sequential([
    keras.layers.Dense(3, input_shape=(2,), activation='sigmoid'),  # Hidden
        layer with 3 neurons
    keras.layers.Dense(1, activation='sigmoid')                     # Output
        layer
])
```

```
# Compile the model
model.compile(optimizer='adam',
              loss='binary_crossentropy',
              metrics=['accuracy'])

# Train the model
model.fit(X, Y, epochs=1000, verbose=0)

# Network test with new data
new_input = np.array([[0.5, 0.5]], dtype=np.float32)
predicted_output = model.predict(new_input)[0][0]

# Display output
print(predicted_output)
```

feedforwardnet In this example, we first define the training data (inputs and outputs), then we create a feedforward neural network using the function. Then we train the network with the training data using the function. Finally, we test a new example using the trained network and get the output.

9.3 Radial Basis Networks

Radial Basis Network (RBF) is a feedforward neural network (FF) that uses a Radial Basis Function (RBF) as the activation function instead of a Logistic Function. The following figure shows a block diagram of this type of neural network.

The logistic function maps some arbitrary values to a range of 0–1 to answer a question (yes or no). This type of neural network is suitable for classification and decision-making systems, but does not perform well for continuous values. Radial basis functions answer the question "How far are we from the goal?" This makes these neural networks suitable for function estimation and machine control. In short, radial basis neural networks are a type of feedforward neural network with different activation functions and properties.

The sample code for building a radial branching neural network (RBF) is as follows:

MATLAB

```
% Definition of training data
x_train = [ 0 0; 1 0; 0 1; 1 1 ] ;
y_train = [ 0; 1; 1; 0 ] ;
% Number of radial neurons
num_neurons = 4;
% Radial neural network training
net = newrb ( x_train',y_train', 0, 1, num_neurons ) ;
% Network testing with test data
x_test = [ 0.5 0.5 ; 0.2 0.8 ] ;
y_pred = sim ( net, x_test');
% Show results
disp(' Predictions: ');
disp(y_pred' ) ;
```

Python

```
import numpy as np

# Definition of training data
x_train = np.array([[0, 0], [1, 0], [0, 1], [1, 1]])  # Inputs
y_train = np.array([[0], [1], [1], [0]])               # Outputs

# Number of radial neurons
num_neurons = 4

# Radial neural network training
class RBFCosmos:
    def __init__(self, num_neurons):
        self.num_neurons = num_neurons
        self.centers = None
        self.sigmas = None
        self.weights = None

    def fit(self, X, y):
        # Initialize centers using k-means clustering
        self.centers, _ = self._kmeans(X, self.num_neurons)

        # Compute sigmas as the average distance between centers
        dists = self._euclidean_distance(self.centers, self.centers)
        self.sigmas = np.mean(np.max(dists, axis=0))

        # Compute the RBF activation matrix
        radial_matrix = self._rbf_activation(X)

        # Train the output layer using linear regression
        self.weights = np.linalg.lstsq(radial_matrix, y, rcond=None)[0]

    def predict(self, X):
        radial_matrix = self._rbf_activation(X)
        return np.dot(radial_matrix, self.weights)

    def _kmeans(self, X, k):
        # Simple k-means implementation for center initialization
        np.random.seed(0)
        centers = X[np.random.choice(X.shape[0], size=k, replace=False)]
        prev_centers = np.zeros_like(centers)
        while not np.allclose(centers, prev_centers):
            prev_centers = centers.copy()
            dists = self._euclidean_distance(X, centers)
            labels = np.argmin(dists, axis=1)
            for i in range(k):
                centers[i] = np.mean(X[labels == i], axis=0)
        return centers, labels

    def _euclidean_distance(self, a, b):
        return np.sqrt(np.sum((a[:, np.newaxis] - b)**2, axis=2))

    def _rbf_activation(self, X):
        dists = self._euclidean_distance(X, self.centers)
        return np.exp(-0.5 * (dists / self.sigmas) ** 2)

# Initialize and train the network
net = RBFCosmos(num_neurons=num_neurons)
net.fit(x_train, y_train)

# Network testing with test data
x_test = np.array([[0.5, 0.5], [0.2, 0.8]])
```

```
y_pred = net.predict(x_test)

# Show results
print('Predictions:')
print(y_pred.reshape(-1))
```

In this code, first, the training data (x_train and y_train) is defined. Then, a radial neural network is created with a specified number of neurons (num_neurons). In this example, the number of neurons is manually set to 4. The network is then trained using the training data. Finally, using the test data (x_test), the neural network makes predictions and displays the results. Note that this is a simple example and may require further adjustments to get the best performance.

9.4 Deep Feed Forward Neural Networks: A Type of Neural Network

Deep Feed Forward Neural Networks (DFF) were introduced to the discussion of neural networks in the early 1990s. These types of neural networks are also feed-forward neural networks, but they have more than one hidden layer.

When training a feed-forward neural network, only a small portion of the error is passed to the previous layer. Therefore, using more layers leads to exponential growth of training time, which makes deep feed-forward neural networks practically useless and impractical. In the early 2000s, approaches were developed that allowed for efficient training of a deep feed-forward neural network (DFF).

Today, these neural networks form the core of modern machine learning systems and serve a similar purpose to that of feed-forward neural networks (FF), but with better results. Here is a sample code to build a deep feed-forward neural network:

MATLAB

```
% Definition of training data
x_train = [0 0; 1 0; 0 1; 1 1];
y_train = [0; 1; 1; 0];
% Definition of neural network architecture
layers = [
    fullyConnectedLayer(10, 'Activation', 'relu')
    fullyConnectedLayer(1, 'Activation', 'sigmoid')
];
% Define training settings
options = trainingOptions('adam', ...
    'MaxEpochs', 100, ...
    'MiniBatchSize', 2, ...
    'InitialLearnRate', 0.01, ...
    'Plots', 'training-progress');
% Building a neural network
net = trainNetwork(x_train', y_train', layers, options);
% Network testing with test data
x_test = [0.5 0.5; 0.2 0.8];
y_pred = predict(net, x_test');
```

```
% Show results
disp('Predictions:');
disp(y_pred');
```

Python

```
import numpy as np
from tensorflow.keras.models import Sequential
from tensorflow.keras.layers import Dense

# Define training data
X_train = np.array([[0, 0], [1, 0], [0, 1], [1, 1]])
y_train = np.array([[0], [1], [1], [0]])

# Define neural network architecture
model = Sequential()  # Create a sequential model
model.add(Dense(10, activation='relu', input_shape=(2,)))  # Add a dense
    layer with 10 units and ReLU activation
model.add(Dense(1, activation='sigmoid'))  # Add a dense layer with 1 unit
    and Sigmoid activation

# Define training settings
model.compile(optimizer='adam',  # Use Adam optimizer
              loss='binary_crossentropy',  # Binary cross-entropy loss
                  function
              metrics=['accuracy'])  # Metrics to evaluate performance

# Train the neural network
model.fit(X_train, y_train, epochs=100, batch_size=2)  # Training process
    without visualization of progress

# Network testing
X_test = np.array([[0.5, 0.5], [0.2, 0.8]])
y_pred = model.predict(X_test)

# Show results
print("Predictions:")
print(y_pred)
```

In this code, first the training data (x_train and y_train) are defined. Then the neural network architecture is specified using two layers (a hidden layer with 10 neurons and an output layer with 1 neuron). Here, ReLU activation functions are used for the hidden layer and sigmoid for the output layer.

The training settings are then specified, and the network is trained using the functions trainNetwork and trainingOptions. Finally, using the test data (x_test), the neural network makes predictions and displays the results. Please note that this is a simple example and may require further adjustments to achieve the best performance.

9.5 Recurrent Neural Network

Recurrent Neural Networks (RNN) introduce a different type of cell called a recurrent cell. The first of these is the Jordan Network. In this type of network, each hidden cell receives its output with a fixed delay (one or more iterations).

Passing state to input nodes, delaying variables, and more are present in recurrent neural networks. This type of neural network is primarily used when context is important, that is, when decisions from previous iterations or instances can affect current instances.

A common example of this type of context is text. In text, a word can only be analyzed in the context of the previous word or sentence. The following is a sample code to build a Recurrent Neural Network (RNN):

MATLAB

```
options = trainingOptions ( 'adam' , ...
    'MaxEpochs' , 100, ...
    'MiniBatchSize' , 1, ...
    'InitialLearnRate' , 0.01 , ...
    'Plots' , 'training-progress' ) ;
% Building a Recurrent Neural Network
net = trainNetwork ( x_train, y_train, layers, options ) ;
% Network testing with test data
x_test = randn ( 5, 1, 1 ) ;
y_pred = predict ( net, x_test ) ;
% Show results
disp ( 'Predictions:' ) ;
disp ( y_pred ) ;
```

Python

```
import torch
import torch.nn as nn
import torch.optim as optim
import matplotlib.pyplot as plt

# Define training options
max_epochs = 100
mini_batch_size = 1
initial_learning_rate = 0.01

# Define a simple RNN model
class RNN(nn.Module):
    def __init__(self, input_size, hidden_size, output_size):
        super(RNN, self).__init__()
        self.rnn = nn.RNN(input_size, hidden_size, batch_first=True)
        self.fc = nn.Linear(hidden_size, output_size)

    def forward(self, x):
        out, _ = self.rnn(x)
        out = self.fc(out[:, -1, :])
        return out

# Create model instance
input_size = 1  # Input feature dimension
hidden_size = 10  # Hidden layer dimension
output_size = 1  # Output dimension
net = RNN(input_size, hidden_size, output_size)

# Define loss function and optimizer
criterion = nn.MSELoss()
optimizer = optim.Adam(net.parameters(), lr=initial_learning_rate)

# Assume training data (replace with actual data during training)
x_train = torch.randn(100, 5, 1).float()  # [batch_size, sequence_length,
    input_size]
```

```
y_train = torch.randn(100, 1).float()    # [batch_size, output_size]

# List to store loss values for plotting
loss_values = []

# Training loop
for epoch in range(max_epochs):
    # Forward pass
    outputs = net(x_train)
    loss = criterion(outputs, y_train)

    # Backward pass and optimization
    optimizer.zero_grad()
    loss.backward()
    optimizer.step()

    # Store loss value
    loss_values.append(loss.item())

    # Print loss every 10 epochs
    if (epoch + 1) % 10 == 0:
        print(f'Epoch [{epoch+1}/{max_epochs}], Loss: {loss.item():.4f}')

# Test data
x_test = torch.randn(1, 5, 1).float()  # [batch_size, sequence_length,
    input_size]

# Set model to evaluation mode
net.eval()
with torch.no_grad():
    y_pred = net(x_test)

# Display results
print('Predictions:')
print(y_pred.numpy())

# Plot loss over epochs
plt.plot(range(1, max_epochs + 1), loss_values)
plt.xlabel('Epochs')
plt.ylabel('Loss')
plt.title('Training Loss vs. Epochs')
plt.show()
```

In this code, first, the training data (x_train and y_train) are defined. Then, the architecture of a recurrent neural network with an LSTM (Long Short-Term Memory) layer and an output layer in the form of a sequence is specified.

The training settings are also determined using the function trainingOptions and using the functions trainNetwork and predict, the neural network is trained, and then prediction is performed with the test data.

9.6 Auto Encoder Neural Network

Auto Encoder (AE) neural networks are used for classification, clustering, and feature compression. When a feedforward neural network is trained for classification, it is given X examples in Y categories as input and expects one of the Y cells to be activated. This is called supervised learning. On the other hand, autoencoder neural networks can be trained unsupervised.

Given the structure of these networks (where the number of hidden layers is smaller than the number of input cells and the number of output cells is equal to the input cells) and that the AE is trained in a way that the output is as close to the input as possible, the autoencoder neural network is forced to generalize the data and look for common patterns. The following is a sample code for building an autoencoder neural network:

MATLAB

```
% Definition of training data
x_train = randn(100, 10); % Input data

% Definition of self-encoding neural network architecture
encoder_neurons = 5;

% Training the self-encoding neural network
autoencoder = trainAutoencoder(x_train', encoder_neurons);

% Using the trained network to decode data
x_decoded = predict(autoencoder, x_train');

% Display results
disp('Input data:');
disp(x_train);
disp('Decoded data (transposed):');
disp(x_decoded'); % Transposed to display correctly
```

Python

```
import numpy as np
from sklearn.decomposition import MiniBatchDictionaryLearning

# Define training data
x_train = np.random.randn(100, 10)  # 100 samples with 10 features

# Define the number of components (encoder neurons)
encoder_neurons = 5

# Initialize the model
autoencoder = MiniBatchDictionaryLearning(n_components=encoder_neurons)

# Train the model on the correct data shape (samples as rows)
autoencoder.fit(x_train)

# Compute the sparse codes (encoded data)
sparse_codes = autoencoder.transform(x_train)

# Reconstruct the data (decode)
x_decoded = sparse_codes @ autoencoder.components_  # Matrix multiplication

# Display results
print("Input Data:")
print(x_train)

print("Decoded Data:")
print(x_decoded)
```

In this code, first the training data (x_train) is defined. Then, using the function trainAutoencoder, a self-encoding neural network with the specified number of hidden layer neurons (here encoder_neurons) is trained.

Then, using the function predict, the training data is sent to the trained neural network and the decoded data x_decodedis stored in the variable.

9.7 Denoising AutoEncoder Neural Network

While autoencoder networks are interesting, they sometimes just adapt to the input data instead of finding the most robust feature. Denoising AutoEncoder (DAE) neural network adds a little noise to the input cell.

This forces the denoising autoencoder to reconstruct the output from a noisy input, making it more general and selecting more common features. Here is a sample code to build a denoising autoencoder:

MATLAB

```
% Definition of training data
x_train = randn(100, 10); % Input data

% Creating noise in the training data
noise_factor = 0.2;
x_train_noisy = x_train + noise_factor * randn(size(x_train));

% Definition of the architecture of the autoencoder
encoder_neurons = 5;

% Training the denoising autoencoder
autoencoder = trainAutoencoder(x_train_noisy', encoder_neurons, ...
    'L2WeightRegularization', 0.001, ...
    'SparsityRegularization', 1, ...
    'SparsityProportion', 0.1, ...
    'MaxEpochs', 1000);

% Using the trained network to decode data
x_decoded = predict(autoencoder, x_train_noisy');

% Show results
disp('Input data:');
disp(x_train);
disp('Decoded data:');
disp(x_decoded);
```

Python

```
import numpy as np
import tensorflow as tf
from tensorflow.keras import layers, Model
from tensorflow.keras.regularizers import l2

# Generate training data
x_train = np.random.randn(100, 10)  # Input data

# Add noise
noise_factor = 0.2
x_train_noisy = x_train + noise_factor * np.random.randn(*x_train.shape)

```

```
# Define the autoencoder architecture
input_dim = x_train.shape[1]
encoder_neurons = 5

# Custom sparsity constraint function using KL divergence
def sparse_kl_divergence(encoder_output, sparsity_target=0.1):
    kl_div = tf.keras.losses.KLDivergence()
    return kl_div(sparsity_target, encoder_output)

# Build the denoising autoencoder
# Define the encoder
encoder_input = layers.Input(shape=(input_dim,))
encoder = layers.Dense(units=encoder_neurons, activation='sigmoid',
    activity_regularizer=l2(0.001))(encoder_input)

# Define the decoder
decoder = layers.Dense(units=input_dim, activation='sigmoid')(encoder)

# Define the autoencoder model
autoencoder = Model(inputs=encoder_input, outputs=decoder)

# Compile the model
autoencoder.compile(optimizer='adam', loss='mse')

# Train the autoencoder
autoencoder.fit(x_train_noisy, x_train, epochs=1000, batch_size=32, shuffle=
    True)

# Use the trained model to decode data
x_decoded = autoencoder.predict(x_train_noisy)

# Display the results
print("Input data:")
print(x_train)
print("Decoded data:")
print(x_decoded)
```

In this code, first, the training data (x_train) is defined. Then, by adding a noise level to the training data (x_train_noisy), the model inputs are generated in a modified form.

Then, using the function, a denoising self-encoding neural network is trained, trainAutoencoder with the specified number of hidden layer neurons (here). In this example, the methods of adding weights to the model ( ) and adjusting the sparsity properties are also used. Then, using the function, the training data is sent to the trained neural network, and the decoded data is stored in the variable encoder_neuronsL2WeightRegularizationpredictx_decoded.

9.8 Variational Auto Encoder Neural Network

The Variational Auto Encoder (VAE) neural network compresses probabilities rather than features, compared to the autoencoder. Despite the small changes that have occurred between the two neural networks, each of these types of artificial neural networks answers a different question. The autoencoder answers the question "How can we generalize the data?", while the variational autoencoder answers the question "How strong is the connection between two events? Should the error be distributed between the two events, or are they completely independent?"

Unfortunately, there is currently no specific function in MATLAB for training a Variational Autoencoder (VAE). However, you can implement your variational autoencoder from programming packages. One way to implement a VAE in MATLAB is to use the functions of the Keras library in Python and interface it with MATLAB. The following is a sample code that uses this approach:

MATLAB

```
% Define training data
x_train = randn(100, 10);

% Transfer training data to Python
py_x_train = py.numpy.array(x_train);

% Create a VAE variable in Python
py_vae = py.VAE_module.create_vae(py_x_train);

% Train VAE
py.VAE_module.train_vae(py_vae, py_x_train);

% Get results from Python
py_decoded_data = py.VAE_module.decode_data(py_vae, py_x_train);
decoded_data = double(py.array.array('d', py_decoded_data));

% Display results
disp('Input data:');
disp(x_train);
disp('Decrypted data:');
disp(decoded_data);
```

This example assumes that you VAE_modulehave a Python library with the appropriate functions for creating and training VAEs. These functions should use well-known libraries such as TensorFlowor to implement VAEs.PyTorch If you want to learn more about this method or a more detailed implementation of VAE in MATLAB, you may want to refer to the related information resources for the Python programming language and Machine Learning libraries. To learn more about various Python libraries such as TensorFlow and Python, read the related ready-made files in this field.

9.9 Long/Short Term Memory

Long/Short Term Memory (LSTM) introduces a new type of memory cell. This cell can process data when it has a time gap (or time delay). A feedforward neural network can process text by remembering the previous ten words. While an LSTM can process video frames by remembering what happened in the previous frames.

LSTM networks are widely used for speech recognition and handwriting recognition. Memory cells are essentially a combination of a pair of elements called gates. These elements are recursive and control how information is remembered and forgotten. Here is a sample code to build a neural network with long short-term memory (LSTM):

MATLAB

```
% Definition of training data
x_train = randn(100, 10); % Input data
y_train = randn(100, 1); % Outputs

% Definition of neural network architecture with long short-term memory
numHiddenUnits = 10;
layers = [
    sequenceInputLayer(size(x_train, 2)) % Sequence input layer
    lstmLayer(numHiddenUnits, 'OutputMode', 'last') % LSTM layer
    fullyConnectedLayer(1) % Fully connected layer
    regressionLayer % Regression layer
];

% Define training settings
options = trainingOptions('adam', ...
    'MaxEpochs', 50, ... % Maximum number of epochs
    'MiniBatchSize', 16, ... % Mini-batch size
    'InitialLearnRate', 0.01, ... % Initial learning rate
    'Plots', 'training progress'); % Plot training progress

% Building a neural network with long short-term memory
net = trainNetwork(x_train, y_train, layers, options);

% Network testing with test data
x_test = randn(10, 10); % Test data
y_pred = predict(net, x_test); % Predictions

% Display results
disp('Predictions:');
disp(y_pred);
```

Python

```
import numpy as np
import tensorflow as tf

# Define training data
x_train = np.random.randn(100, 10)  # Input data
y_train = np.random.randn(100, 1)  # Outputs

# Define test data
x_test = np.random.randn(10, 10)   # Test data

# Build LSTM neural network
numHiddenUnits = 10  # Number of LSTM units
model = tf.keras.Sequential([
    tf.keras.layers.LSTM(numHiddenUnits, input_shape=(1, 10),
        return_sequences=False),  # LSTM layer
    tf.keras.layers.Dense(1, activation='linear')  # Fully connected layer
        with linear activation (regression)
])

# Define training settings
```

```
optimizer = tf.keras.optimizers.Adam(learning_rate=0.01)
model.compile(optimizer=optimizer, loss='mse')  # Mean squared error loss
    for regression

# Reshape input data to match LSTM input shape (samples, time steps,
    features)
x_train_reshaped = np.expand_dims(x_train, axis=1)  # Shape: (100, 1, 10)
x_test_reshaped = np.expand_dims(x_test, axis=1)    # Shape: (10, 1, 10)

# Train the network
history = model.fit(
    x_train_reshaped,
    y_train,
    batch_size=16,
    epochs=50,
    verbose=1  # Show training progress
)

# Make predictions with the trained model
y_pred = model.predict(x_test_reshaped)
y_pred_flat = y_pred.flatten()  # Flatten the prediction array for display

# Display predictions
print("Predictions:")
print(y_pred_flat)
```

In this code, first, the training data (x_train and y_train) are defined. Then, the architecture of a long short-term memory neural network with an LSTM layer (and a specified number of neurons) and an output layer is defined.

The training settings are also determined using the function trainingOptions and using the functions trainNetwork and predict, the network is trained, and prediction is performed with the test data.

9.10 Gated Recurrent Unit Neural Network

Gated Recurrent Unit (GRU) is a type of LSTM with different gates and periods. This type of neural network seems simple. The lack of an output gate makes it easier to repeat the same output multiple times for the inputs. This type of recurrent neural network is currently used mostly in speech synthesis and music synthesis engines.

Of course, the actual combination of LSTM and GRU is slightly different. Because all the gates of LSTM are combined into one gate called the update gate, and the reset gate is closely tied to the input. GRUs use fewer resources than LSTMs and have a similar effect. Below is a sample code to build a Gated Recurrent Unit (GRU) neural network:

MATLAB

```
% Definition of training data
x_train = randn(100, 10);   % Input data
y_train = randn(100, 1);    % Outputs

% Definition of neural network architecture with GRU recurrent unit
numHiddenUnits = 10;
layers = [
```

```
    sequenceInputLayer(size(x_train, 2))
    gruLayer(numHiddenUnits)
    fullyConnectedLayer(1)
    regressionLayer
];

% Define training settings
options = trainingOptions('adam', ...
    'MaxEpochs', 50, ...
    'MiniBatchSize', 16, ...
    'InitialLearnRate', 0.01, ...
    'Plots', 'training-progress');

% Building a neural network with a recurrent unit
net = trainNetwork(x_train, y_train, layers, options);

% Network testing with test data
x_test = randn(10, 10);
y_pred = predict(net, x_test);

% Show results
disp('Predictions:');
disp(y_pred); % Display predictions
```

Python

```
import tensorflow as tf

# Define training data
x_train = tf.random.normal((100, 10))      # Input data (100 samples, 10 features)
y_train = tf.random.normal((100, 1))       # Output data (100 samples, 1 output)

# Define neural network architecture with GRU recurrent unit
num_hidden_units = 10

# Build the model
inputs = tf.keras.layers.Input(shape=(10,))
x = tf.keras.layers.Reshape((10, 1))(inputs)  # Reshape to (sequence_length, features)
x = tf.keras.layers.GRU(num_hidden_units)(x)  # GRU layer
outputs = tf.keras.layers.Dense(1)(x)         # Fully connected layer for regression
model = tf.keras.Model(inputs=inputs, outputs=outputs)

# Compile the model
model.compile(
    optimizer=tf.keras.optimizers.Adam(learning_rate=0.01),
    loss=tf.keras.losses.MeanSquaredError()
)

# Train the model
model.fit(
    x_train,
    y_train,
    epochs=50,
    batch_size=16,
    verbose=1
)

# Test the network
x_test = tf.random.normal((10, 10))
```

```
y_pred = model.predict(x_test)

# Display predictions
print("Predictions:")
print(y_pred)
```

In this code, the architecture of a neural network with a recurrent unit (gruLayer) is defined. The training settings are also determined using the function trainingOptions. Using the functions trainNetwork and predict, the network is trained, and a prediction is performed on the test data.

9.11 Deep Residual Network

A Deep Residual Network (DRN) is a deep network in which a portion of the input data is passed to subsequent layers. This feature allows these networks to be very deep, up to 300 layers deep, but they are a type of deep convolutional network without explicit delay.

Below is a sample code to build a Deep Residual Network (ResNet). This example uses a simple ResNet with 18 layers:

MATLAB

```
%% Definition of Training Data
x_train = randn(100, 32, 32, 3); % Input data
y_train = randn(100, 1);         % Outputs

%% Define Network Architecture
numClasses = 1;
inputSize = [32 32 3];

layers = [
    imageInputLayer(inputSize);

    convolution2dLayer(7, 64, 'Padding', 3, 'Stride', 2);
    batchNormalizationLayer;
    reluLayer;

    maxPooling2dLayer(3, 'Stride', 2, 'Padding', 1);

    residualBlock(64, 3, 64);
    residualBlock(128, 3, 128);
    residualBlock(256, 3, 256);

    averagePooling2dLayer(7);

    fullyConnectedLayer(numClasses);
    regressionLayer;
];

%% Define Training Settings
options = trainingOptions('adam', ...
    'MaxEpochs', 20, ...
    'MiniBatchSize', 16, ...
    'InitialLearnRate', 0.001, ...
    'Plots', 'training-progress');
```

```

%% Train Network
net = trainNetwork(x_train, y_train, layers, options);

%% Network Testing with Test Data
x_test = randn(10, 32, 32, 3);
y_pred = predict(net, x_test);

%% Show Results
disp('Predictions:');
disp(y_pred);

%% Define residualBlock Function
function layers = residualBlock(numFilters, numBlocks, stride)
    layers = [
        convolution2dLayer(3, numFilters, 'Padding', 'same', 'Stride',
            stride);
        batchNormalizationLayer;
        reluLayer;

        convolution2dLayer(3, numFilters, 'Padding', 'same');
        batchNormalizationLayer;
        additionLayer('Name', 'residual_addition');
        reluLayer;
    ];

    for i = 1:(numBlocks - 1)
        layers = [
            layers,
            convolution2dLayer(3, numFilters, 'Padding', 'same');
            batchNormalizationLayer;
            reluLayer;
            convolution2dLayer(3, numFilters, 'Padding', 'same');
            batchNormalizationLayer;
            additionLayer('Name', 'residual_addition');
            reluLayer;
        ];
    end
end
```

Python

```
mport torch
import torch.nn as nn
import torch.optim as optim
from torch.utils.data import DataLoader, TensorDataset
import numpy as np

# Set random seed for reproducibility
torch.manual_seed(42)

# Define training data
x_train = torch.randn(100, 3, 32, 32).float()  # Input data, shape [
    batch_size, channels, height, width]
y_train = torch.randn(100, 1).float()          # Output data

# Define residual block
class ResidualBlock(nn.Module):
    def __init__(self, in_channels, out_channels, stride=1):
        super(ResidualBlock, self).__init__()
        self.conv1 = nn.Conv2d(in_channels, out_channels, kernel_size=3,
            stride=stride, padding=1, bias=False)
        self.bn1 = nn.BatchNorm2d(out_channels)
```

```
        self.relu = nn.ReLU(inplace=True)
        self.conv2 = nn.Conv2d(out_channels, out_channels, kernel_size=3,
             stride=1, padding=1, bias=False)
        self.bn2 = nn.BatchNorm2d(out_channels)

        # Add a projection layer if the number of input channels is
             different from the output channels or stride != 1
        if stride != 1 or in_channels != out_channels:
            self.downsample = nn.Sequential(
                nn.Conv2d(in_channels, out_channels, kernel_size=1, stride=
                     stride, bias=False),

                nn.BatchNorm2d(out_channels)
            )
        else:
            self.downsample = nn.Identity()

    def forward(self, x):
        identity = x
        out = self.conv1(x)
        out = self.bn1(out)
        out = self.relu(out)
        out = self.conv2(out)
        out = self.bn2(out)
        identity = self.downsample(identity)
        out += identity
        out = self.relu(out)
        return out

# Define network architecture
class ResNet(nn.Module):
    def __init__(self, num_classes=1):
        super(ResNet, self).__init__()
        self.in_channels = 64

        self.conv1 = nn.Conv2d(3, 64, kernel_size=7, stride=2, padding=3,
             bias=False)
        self.bn1 = nn.BatchNorm2d(64)
        self.relu = nn.ReLU(inplace=True)
        self.maxpool = nn.MaxPool2d(kernel_size=3, stride=2, padding=1)

        self.layer1 = self._make_layer(64, 64, blocks=1, stride=1)
        self.layer2 = self._make_layer(64, 128, blocks=1, stride=2)
        self.layer3 = self._make_layer(128, 256, blocks=1, stride=2)

        self.avgpool = nn.AdaptiveAvgPool2d((1, 1))
        self.fc = nn.Linear(256, num_classes)

    def _make_layer(self, in_channels, out_channels, blocks, stride):
        layers = []
        layers.append(ResidualBlock(in_channels, out_channels, stride))
        for _ in range(1, blocks):
            layers.append(ResidualBlock(out_channels, out_channels, stride
                 =1))
        return nn.Sequential(*layers)

    def forward(self, x):
        x = self.conv1(x)
        x = self.bn1(x)
        x = self.relu(x)
        x = self.maxpool(x)

        x = self.layer1(x)
```

```
        x = self.layer2(x)
        x = self.layer3(x)

        x = self.avgpool(x)
        x = torch.flatten(x, 1)
        x = self.fc(x)
        return x

# Initialize the network
net = ResNet(num_classes=1)

# Define loss function and optimizer
criterion = nn.MSELoss()  # For regression tasks
optimizer = optim.Adam(net.parameters(), lr=0.001)

# Define training settings
max_epochs = 20
batch_size = 16

# Convert data to DataLoader
train_dataset = TensorDataset(x_train, y_train)
train_loader = DataLoader(train_dataset, batch_size=batch_size, shuffle=True
    )

# Train the network
for epoch in range(max_epochs):
    net.train()
    running_loss = 0.0
    for i, data in enumerate(train_loader, 0):
        inputs, targets = data

        # Zero the parameter gradients
        optimizer.zero_grad()

        # Forward pass
        outputs = net(inputs)
        loss = criterion(outputs, targets)

        # Backward pass and optimize
        loss.backward()
        optimizer.step()

        running_loss += loss.item()

    # Print training progress
    print(f"Epoch [{epoch+1}/{max_epochs}], Loss: {running_loss/(i+1):.4f}")

# Test the network
x_test = torch.randn(10, 3, 32, 32).float()
net.eval()
with torch.no_grad():
    y_pred = net(x_test)

# Show results
print("Predictions:")
print(y_pred.numpy())
```

In this code, the ResNet architecture is defined using layers convolution2dLayer, batchNormalizationLayer, reluLayer, and. The function is also used to create the remaining blocks in the network. The functions are also used to train and test the network, respectively.additionLayerresidualBlocktrainNetworkpredict.

9.12 Markov Chains

Markov Chains are a concept of graphs in which each edge has a probability. In ancient times, Markov chains were used to construct text; for example, after the word Hello, the word Dear comes with a probability of 0.0053%, and the word You with a probability of 0.03551%.

Markov chain neural networks are used as a concise model to predict or generate sequences based on the dependence on past states. A common approach to this is to use Recurrent Neural Networks (RNNs). Below is a sample code to build a Markov chain RNN model:

MATLAB

```
% Definition of training data
sequenceLength = 100;          % Length of sequences
numSequences = 50;             % Number of sequences
inputSize = 1;                 % Input dimension
X_train = randn(sequenceLength, numSequences, inputSize); % Generate random
    training data

% Define recursive network architecture
numHiddenUnits = 10;           % Number of hidden units
layers = [
    sequenceInputLayer(inputSize) % Input layer
    lstmLayer(numHiddenUnits, 'OutputMode', 'sequence') % LSTM layer
    fullyConnectedLayer(1) % Fully connected layer
    regressionLayer % Regression layer
];

% Define training options
options = trainingOptions('adam', ...
    'MaxEpochs', 50, ...        % Maximum number of training epochs
    'MiniBatchSize', 16, ...    % Mini-batch size
    'InitialLearnRate', 0.01, ... % Initial learning rate
    'Plots', 'training-progress'); % Plot training progress

% Train the recursive network
net = trainNetwork(X_train, X_train, layers, options); % Train the network
    with training data

% Test the network with test data
X_test = randn(sequenceLength, 10, inputSize); % Generate random test data
Y_pred = predict(net, X_test); % Prediction

% Display results
disp('Predictions:');
disp(Y_pred);
```

Python

```
import numpy as np
import tensorflow as tf
from tensorflow.keras.models import Sequential
from tensorflow.keras.layers import LSTM, Dense, TimeDistributed
import matplotlib.pyplot as plt

# Define training data
sequence_length = 100           # Sequence length
```

```
num_sequences = 50          # Number of sequences
input_size = 1              # Input dimension

# Generate random training data (similar to MATLAB's randn)
X_train = np.random.randn(num_sequences, sequence_length, input_size)

# Define the recurrent network architecture
num_hidden_units = 10       # Number of LSTM internal units
model = Sequential([
    tf.keras.layers.InputLayer(input_shape=(sequence_length, input_size)),
    LSTM(num_hidden_units, return_sequences=True),  # LSTM layer, outputs all time steps
    TimeDistributed(Dense(1)),                      # Fully connected layer (outputs per time step)
])

# Define training options
# Use the Adam optimizer, mean squared error (MSE) loss function, and monitor training progress
optimizer = tf.keras.optimizers.Adam(learning_rate=0.01)
model.compile(optimizer=optimizer, loss='mse')

# Display model structure
model.summary()

# Train the network
history = model.fit(
    X_train, X_train,           # Input and target are both X_train (similar to an autoencoder)
    epochs=50,                  # Maximum number of training epochs
    batch_size=16,              # Batch size
    verbose=1,                  # Display training progress
)

# Plot training loss
plt.plot(history.history['loss'])
plt.title('Training Loss')
plt.xlabel('Epoch')
plt.ylabel('Loss (MSE)')
plt.show()

# Test the network
num_test_sequences = 10
X_test = np.random.randn(num_test_sequences, sequence_length, input_size)  # Generate test data
Y_pred = model.predict(X_test)  # Prediction

# Display results
print("Predictions:")
print(Y_pred)
```

In this code, a recurrent model with a single LSTM layer is defined. This model uses the training data to predict or generate sequences based on a Markov chain. Note that this code is a simple example. You can modify the model with different settings and architectures to suit your needs.

To have an attractive and audience-friendly presentation on the topic of Markov chains, read the prepared file available on this topic. • Markov chain PowerPoint—Click.

9.13 Hopfield Neural Networks

Hopfield Networks (HN) are a type of neural network that is trained on a limited set of examples, so they respond to a known example with a similar example. Before training, each cell acts as an input cell, during training as a hidden cell, and when in use, as an output cell. A Hopfield network attempts to build trained examples, and these networks are used to denoise and retrain inputs. If these networks are given half of a learned image or sequence, they will return the full example.

In general, Hopfield networks are a type of neural network from the recurrent network class that is used to solve optimization and chaotic memory problems. Below is a sample code to build a Hopfield network:

MATLAB

```
% Define training patterns (each pattern is a column vector)
patterns = [+1, -1, +1, -1;   % First pattern (transposed to a column vector
    )
            -1, -1, +1, +1]'; % Second pattern, transposed to a column
                vector

% Define Hopfield network
numNeurons = size(patterns, 1); % Number of neurons = pattern dimension (
    number of rows)
net = newhop(patterns);            % Create network

% Test network memory
inputPattern = [+1, -1, +1, +1]'; % Input pattern transposed to a column
    vector (4x1)
outputPattern = sim(net, {1, 1}, [], [], inputPattern); % Correct parameter
    order

% Display results
disp('Input pattern (transposed):');
disp(inputPattern');

disp('Output pattern:');
disp(cell2mat(outputPattern)');
```

Python

```
import numpy as np

# Definition of training data (patterns)
patterns = np.array([
    [+1, -1, +1, -1],   # First row pattern
    [-1, -1, +1, +1]    # Second row pattern
])

# Definition of Hopfield network architecture
num_neurons = patterns.shape[1]  # Calculate the number of neurons

# Calculate the weight matrix for Hopfield network
# Hopfield weights are calculated using Hebb's rule
# W = (patterns.T @ patterns) / num_neurons
W = (patterns.T @ patterns) / num_neurons
np.fill_diagonal(W, 0)  # Set diagonal elements to zero to avoid self-
    feedback

# Network memory test
```

```python
input_pattern = np.array([+1, -1, +1, +1])  # Define the input pattern
output_pattern = np.copy(input_pattern)  # Initialize output pattern

# Simulate Hopfield network until convergence
max_iterations = 10  # Maximum number of iterations to avoid infinite loops
for _ in range(max_iterations):
    # Calculate the next state using the weight matrix
    next_output = np.sign(W @ output_pattern)

    # Check for convergence (no change in state)
    if np.array_equal(next_output, output_pattern):
        break

    output_pattern = next_output

# Show results
print("Input pattern:")
print(input_pattern)

print("Output pattern:")
print(output_pattern)
```

In this code, patterns there is a set of patterns (memory patterns) that the network is trained on. Then, a Hopfield network is created using the function newhop. Finally, the input pattern is tested using the function simand the output pattern is displayed.

Please note that Hopfield networks in practice are usually used to solve specific problems, such as chaotic memory and require specific input and training settings. This code sample is provided as a simple example of a Hopfield network, and you can modify it to suit your needs.

9.14 Boltzmann Machines

Boltzmann Machines (BM) are very similar to Hopfield networks in that some cells are marked as inputs and remain hidden. Input cells become output cells as soon as the hidden cells update their state (during training, the Boltzmann machine/Hopfield network updates cells one by one, not in parallel).

In general, Boltzmann machines are a type of probabilistic graphical model used in the field of machine learning. Below is a sample code to build a Boltzmann machine:

MATLAB

```matlab
% Define the logistic function
logistic = @(x) 1 ./ (1 + exp(-x));

% Define training data
trainingData = [ ...
    +1, -1, +1, -1; ...
    -1, -1, +1, +1 ...
];

% Determine the number of neurons
numVisible = size(trainingData, 1);
numHidden = size(trainingData, 2);
```

```

% Initialize Boltzmann Machine parameters
weights = randn(numVisible, numHidden);
visibleBias = randn(numVisible, 1);
hiddenBias = randn(numHidden, 1);

% Define training parameters
learningRate = 0.1;
numEpochs = 100;

% Training loop
for epoch = 1:numEpochs
    % Compute hidden layer probabilities
    hiddenProb = logistic(weights' * trainingData + repmat(hiddenBias, 1,
        size(trainingData, 2)));

    % Hidden layer sampling
    hiddenStates = hiddenProb > rand(numHidden, size(trainingData, 2));

    % Compute visible layer probabilities
    visibleProb = logistic(weights * hiddenStates + repmat(visibleBias, 1,
        size(trainingData, 2)));

    % Visible layer sampling
    visibleStates = visibleProb > rand(numVisible, size(trainingData, 2));

    % Update parameters
    weights = weights + learningRate * (trainingData * hiddenStates' -
        visibleStates * hiddenProb');
    visibleBias = visibleBias + learningRate * sum(trainingData -
        visibleStates, 2);
    hiddenBias = hiddenBias + learningRate * sum(hiddenProb - hiddenStates,
        2);
end

% Display weights and biases
disp('Weights:');
disp(weights);
disp('Visible neuron bias:');
disp(visibleBias);
disp('Hidden neuron bias:');
disp(hiddenBias);
```

Python

```python
import numpy as np

# Define the logistic function
def logistic(x):
    return 1 / (1 + np.exp(-x))

# Define training data
trainingData = np.array([
    [+1, -1, +1, -1],
    [-1, -1, +1, +1]
])

# Determine the number of neurons
numVisible, numHidden = trainingData.shape

# Initialize Boltzmann Machine parameters
weights = np.random.randn(numVisible, numHidden)
visibleBias = np.random.randn(numVisible, 1)
```

```
hiddenBias = np.random.randn(numHidden, 1)

# Define training parameters
learningRate = 0.1
numEpochs = 100

# Training loop
for epoch in range(numEpochs):
    # Compute hidden layer probabilities
    hiddenInput = np.dot(weights.T, trainingData) + np.tile(hiddenBias, (1, trainingData.shape[1]))
    hiddenProb = logistic(hiddenInput)

    # Hidden layer sampling
    hiddenStates = hiddenProb > np.random.rand(numHidden, trainingData.shape[1])

    # Compute visible layer probabilities
    visibleInput = np.dot(weights, hiddenStates) + np.tile(visibleBias, (1, hiddenStates.shape[1]))
    visibleProb = logistic(visibleInput)

    # Visible layer sampling
    visibleStates = visibleProb > np.random.rand(numVisible, hiddenStates.shape[1])

    # Update parameters
    positive_associations = np.dot(trainingData, hiddenStates.T)
    negative_associations = np.dot(visibleStates, hiddenProb.T)
    weights += learningRate * (positive_associations - negative_associations)

    visibleBias += learningRate * np.sum(trainingData - visibleStates, axis=1, keepdims=True)
    hiddenBias += learningRate * np.sum(hiddenProb - hiddenStates, axis=1, keepdims=True)

# Display weights and biases
print("Weights:")
print(weights)
print("Visible neuron bias:")
print(visibleBias)
print("Hidden neuron bias:")
print(hiddenBias)
```

In this code, a Boltzmann machine is created with an arbitrary number of visible and hidden neurons. The probabilities of the hidden and visible neurons are calculated with the logit function, and then these probabilities are used for sampling. The Boltzmann machine is updated in each epoch using the Gibbs variable learning algorithm.

9.15 Restricted Boltzmann Machine

The Restricted Boltzmann Machine (RBM) is similar in structure to the BM. However, due to its restriction, it can only be trained using backpropagation, as the feed-forward (with the only difference being that the data passed through backpropagation is fed back to the input layer once).

A restricted Boltzmann machine is a type of probabilistic graph model used in machine learning and dimensionality reduction tools. Below is a sample MATLAB code to build an RBM:

MATLAB

```
% Definition of training data
data = [+1, -1, +1, -1; ...
        -1, -1, +1, +1];

% Number of visible and hidden neurons
numVisible = size(data, 1);
numHidden = 2;

% Parameters of the restricted Boltzmann machine
weights = randn(numVisible, numHidden);
visibleBias = randn(numVisible, 1);
hiddenBias = randn(numHidden, 1);

% Number of training epochs and learning rate
numEpochs = 1000;
learningRate = 0.1;

% Training a restricted Boltzmann machine
for epoch = 1:numEpochs
    % Calculate the probabilities of hidden neurons
    hiddenProb = 1 ./ (1 + exp(- (weights' * data + repmat(hiddenBias, 1,
        size(data, 2)))));

    % Sampling of hidden neurons
    hiddenStates = hiddenProb > rand(numHidden, size(data, 2));

    % Calculate the probabilities of visible neurons
    visibleProb = 1 ./ (1 + exp(-(weights * hiddenStates + repmat(
        visibleBias, 1, size(data, 2)))));

    % Sampling of Visible Neurons
    visibleStates = visibleProb > rand(numVisible, size(data, 2));

    % Update weights and biases
    deltaWeights = learningRate * (data * hiddenStates' - visibleStates *
        hiddenProb');
    deltaVisibleBias = learningRate * sum(data - visibleStates, 2);
    deltaHiddenBias = learningRate * sum(hiddenProb - hiddenStates, 2);

    weights = weights + deltaWeights;
    visibleBias = visibleBias + deltaVisibleBias;
    hiddenBias = hiddenBias + deltaHiddenBias;
end

% Display weights and biases
disp('Weights:');
```

```
disp(weights);
disp('Visible neuron bias:');
disp(visibleBias);
disp('Hidden neuron bias:');
disp(hiddenBias);
```

Python

```
import numpy as np

# Define training data
data = np.array([[1, -1, 1, -1],
                 [-1, -1, 1, 1]])

# Number of neurons in the visible and hidden layers
num_visible = data.shape[0]
num_hidden = 2

# Initialize RBM parameters
weights = np.random.randn(num_visible, num_hidden)
visible_bias = np.random.randn(num_visible, 1)
hidden_bias = np.random.randn(num_hidden, 1)

# Training parameters
num_epochs = 1000
learning_rate = 0.1

# Train the RBM
for epoch in range(num_epochs):
    # Calculate hidden layer probabilities (Correction 1: Removed np.newaxis)
    hidden_input = np.dot(weights.T, data) + hidden_bias  # Original error: hidden_bias[:, np.newaxis]
    hidden_prob = 1 / (1 + np.exp(-hidden_input))

    # Sample hidden layer states
    hidden_states = hidden_prob > np.random.rand(num_hidden, data.shape[1])

    # Calculate visible layer probabilities (Correction 2: Removed np.newaxis)
    visible_input = np.dot(weights, hidden_states) + visible_bias  # Original error: visible_bias[:, np.newaxis]
    visible_prob = 1 / (1 + np.exp(-visible_input))

    # Sample visible layer states
    visible_states = visible_prob > np.random.rand(num_visible, data.shape[1])

    # Update parameters (Corrected dimension matching)
    delta_weights = learning_rate * (np.dot(data, hidden_states.T) - np.dot(visible_states, hidden_prob.T))
    delta_visible_bias = learning_rate * np.sum(data - visible_states, axis=1, keepdims=True)
    delta_hidden_bias = learning_rate * np.sum(hidden_prob - hidden_states, axis=1, keepdims=True)
    weights += delta_weights
    visible_bias += delta_visible_bias
    hidden_bias += delta_hidden_bias

# Output results
print("Weights:")
print(weights)
print("Visible Bias:")
```

```
print(visible_bias)
print("Hidden Bias:")
print(hidden_bias)
```

In this code, a restricted Boltzmann machine is created with a specified number of visible and hidden neurons. The machine is updated with the Gibbs variational training algorithm.

9.16 Deep Convolutional Inverse Graphics Network is a Type of Neural Network

Deep Convolutional Inverse Graphics Network (DCIGN) is a type of deep neural network architecture used to solve inverse graphics problems in fields such as machine vision and image processing. These networks are mainly used in image interpretation and reconstruction problems, 3D structure interpretation, and artificial image generation.

The term "inverse graphics" refers to the attempt to recover and predict the original features or details of the input image from points such as areas, 3D, directions, and other characteristics. In the DCIGN architecture, a deep neural network with convolutional layers and pooling layers is used to extract image features from the input. Then, layers called "inverse graphics" reconstruct the original features of the image.

Typically, these networks are trained on labeled data to learn the inverse graph map of features. DCIGN can be useful in tasks such as image retrieval, interpreting the 3D structure of objects, learning image representations, or generating synthetic images.

To provide sample MATLAB code for a Deep Convolutional Inverse Graph Network (DCIGN), different tasks and specific datasets need to be defined. But you can generically create a DCIGN architecture using convolutional and inverse graph functions and layers. Here is a simple template for a DCIGN:

MATLAB

```
% Network architecture definition
layers = [
    imageInputLayer([64 64 3])  % A 64x64 input with 3 channels for RGB
         images
    convolution2dLayer(3, 64, 'Padding', 'same')
    reluLayer
    maxPooling2dLayer(2, 'Stride', 2)
    convolution2dLayer(3, 128, 'Padding', 'same')
    reluLayer
    maxPooling2dLayer(2, 'Stride', 2)

    % Add inverse graphic layers to restore features
    transposedConv2dLayer(4, 128, 'Stride', 2, 'Cropping', 'same')
    reluLayer
    transposedConv2dLayer(4, 64, 'Stride', 2, 'Cropping', 'same')
    reluLayer
```

```
    transposedConv2dLayer(4, 3, 'Stride', 2, 'Cropping', 'same') % Output
        with 3 channels for RGB image
    regressionLayer
];

% Training settings
options = trainingOptions('adam', ...
    'MaxEpochs', 20, ...
    'MiniBatchSize', 16, ...
    'InitialLearnRate', 0.001, ...
    'Plots', 'training-progress');

% Network construction and training
net = trainNetwork(trainingData, targetData, layers, options);
```

Python

```
import torch
import torch.nn as nn
import torch.optim as optim
from torch.utils.data import DataLoader
from torchvision import transforms, datasets

# Define the network architecture
class ImageReconstructionNet(nn.Module):
    def __init__(self):
        super(ImageReconstructionNet, self).__init__()
        self.encoder = nn.Sequential(
            nn.Conv2d(3, 64, kernel_size=3, padding=1),  # 64x64x3 -> 64
                x64x64
            nn.ReLU(),
            nn.MaxPool2d(2, stride=2),                    # 64x64x64 -> 32
                x32x64
            nn.Conv2d(64, 128, kernel_size=3, padding=1), # 32x32x64 -> 32
                x32x128
            nn.ReLU(),
            nn.MaxPool2d(2, stride=2)                     # 32x32x128 -> 16
                x16x128
        )

        self.decoder = nn.Sequential(
            nn.ConvTranspose2d(128, 128, kernel_size=4, stride=2, padding=1)
                , # 16x16x128 -> 32x32x128
            nn.ReLU(),
            nn.ConvTranspose2d(128, 64, kernel_size=4, stride=2, padding=1),
                # 32x32x128 -> 64x64x64
            nn.ReLU(),
            nn.ConvTranspose2d(64, 3, kernel_size=4, stride=2, padding=1)
                # 64x64x64 -> 128x128x3
        )
    def forward(self, x):
        x = self.encoder(x)
        x = self.decoder(x)
        return x

# Training settings
device = torch.device("cuda" if torch.cuda.is_available() else "cpu")
net = ImageReconstructionNet().to(device)
criterion = nn.MSELoss()  # Mean Squared Error loss for regression problems
optimizer = optim.Adam(net.parameters(), lr=0.001)

# Assume trainingData and targetData are correctly loaded as PyTorch tensors
# The following is example code (assuming image data is used)
```

```
transform = transforms.Compose([
    transforms.Resize((64, 64)),
    transforms.ToTensor()
])

# Assume the dataset is stored in the path data_dir
data_dir = './data'
train_dataset = datasets.ImageFolder(data_dir, transform=transform)
train_loader = DataLoader(train_dataset, batch_size=16, shuffle=True)

# Train the network
num_epochs = 20
for epoch in range(num_epochs):
    net.train()
    running_loss = 0.0
    for inputs, _ in train_loader:  # Assume target data is the same as
         input data (autoencoder task)
        inputs = inputs.to(device)
        targets = inputs  # Autoencoder task, input and target are the same

        optimizer.zero_grad()
        outputs = net(inputs)
        loss = criterion(outputs, targets)
        loss.backward()
        optimizer.step()

        running_loss += loss.item()
    print(f"Epoch {epoch+1}/{num_epochs}, Loss: {running_loss/len(
         train_loader):.4f}")

print('Training completed!')
```

Note that this code is a simple template and requires more fine-tuning and adaptation to the dataset being used. To use this template, you need to substitute your training and target data trainingData and targetData adjust the layers and network parameters according to the characteristics and size of your dataset.

9.17 Generative Adversarial Networks

Generative Adversarial Networks (GANs) are a type of deep neural network architecture that simultaneously collaborates or competes with two networks, called a generator and a discriminator. The two networks engage in a mutual learning process so that the generator can produce synthetic data that appears to be more similar to real data.

The generator in a GAN tries to produce data that looks like real data, while the discriminator tries to distinguish between real data and data generated by the generator. This training process improves both networks and results in the production of high-quality synthetic data.

One of the most popular applications of GANs is in the generation of realistic images and videos. These networks are used in the fields of visual arts, synthetic image generation, training data generation for neural networks, and even image correction and refinement. So far, GANs have enjoyed great success in various fields, but they have also been plagued by challenges and issues such as training stability and instability attacks.

Due to the length and non-primitive nature of the GAN code, we cannot provide a complete GAN code example. However, you can do a simple GAN implementation. Here is a simple GAN template:

MATLAB

```
% Definition of generator
generator = [
    fullyConnectedLayer(256)
    reluLayer
    fullyConnectedLayer(28*28)
    tanhLayer
    reshapeLayer([28 28 1])
];

% Definition of discriminator
discriminator = [
    imageInputLayer([28 28 1])
    fullyConnectedLayer(256)
    leakyReluLayer(0.01)
    fullyConnectedLayer(1)
    sigmoidLayer
];

% Define GAN settings
options = trainingOptions('adam', ...
    'MaxEpochs', 100, ...
    'MiniBatchSize', 128, ...
    'Verbose', true, ...
    'Plots', 'training-progress');

gan = ganNetwork(generator, discriminator, options);

% Training data
data = imageDatastore('path_to_dataset', 'IncludeSubfolders', true, '
    LabelSource', 'foldernames');

% GAN training
trainedGAN = trainNetwork(zeros(28, 28, 1, 128), [], gan, options);
```

Python

```
import tensorflow as tf
from tensorflow.keras import layers, models
import matplotlib.pyplot as plt

# Define the generator
def build_generator(latent_dim):
    model = models.Sequential([
        layers.Dense(256, input_dim=latent_dim, activation='relu'),
        layers.Dense(28 * 28, activation='tanh'),
        layers.Reshape((28, 28, 1))
    ])
    return model

# Define the discriminator
def build_discriminator():
    model = models.Sequential([
        layers.InputLayer(input_shape=(28, 28, 1)),
```

```
        layers.Flatten(),
        layers.Dense(256),
        layers.LeakyReLU(alpha=0.01),
        layers.Dense(1, activation='sigmoid')
    ])
    return model

# Main program
if __name__ == "__main__":
    # Dataset path
    dataset_path = 'path_to_dataset'

    # Load the dataset
    (train_images, _), (_, _) = tf.keras.datasets.mnist.load_data()
    train_images = train_images.reshape(train_images.shape[0], 28, 28, 1).
        astype('float32')
    train_images = (train_images - 127.5) / 127.5  # Normalize to [-1, 1]

    BUFFER_SIZE = 60000  # Dataset size
    BATCH_SIZE = 128
    train_dataset = tf.data.Dataset.from_tensor_slices(train_images).shuffle
        (BUFFER_SIZE).batch(BATCH_SIZE)

    # Set hyperparameters
    latent_dim = 100
    epochs = 100

    # Create generator and discriminator
    generator = build_generator(latent_dim)
    discriminator = build_discriminator()

    # Define optimizers
    cross_entropy = tf.keras.losses.BinaryCrossentropy(from_logits=False)
    generator_optimizer = tf.keras.optimizers.Adam(learning_rate=0.0002,
        beta_1=0.5)
    discriminator_optimizer = tf.keras.optimizers.Adam(learning_rate=0.0002,
        beta_1=0.5)

    # Define generator loss function
    def generator_loss(fake_output):
        return cross_entropy(tf.ones_like(fake_output), fake_output)

    # Define discriminator loss function
    def discriminator_loss(real_output, fake_output):
        real_loss = cross_entropy(tf.ones_like(real_output), real_output)
        fake_loss = cross_entropy(tf.zeros_like(fake_output), fake_output)
        total_loss = real_loss + fake_loss
        return total_loss

    # Generate fixed noise for visualization
    random_vector_for_generation = tf.random.normal(shape=[16, latent_dim])

    # Define training step
    @tf.function
    def train_step(images):
        noise = tf.random.normal([BATCH_SIZE, latent_dim])

        with tf.GradientTape() as gen_tape, tf.GradientTape() as disc_tape:
            generated_images = generator(noise, training=True)

            real_output = discriminator(images, training=True)
            fake_output = discriminator(generated_images, training=True)

```

```
            gen_loss = generator_loss(fake_output)
            disc_loss = discriminator_loss(real_output, fake_output)

        gradients_of_generator = gen_tape.gradient(gen_loss, generator.
            trainable_variables)
        gradients_of_discriminator = disc_tape.gradient(disc_loss,
            discriminator.trainable_variables)

        generator_optimizer.apply_gradients(zip(gradients_of_generator,
            generator.trainable_variables))
        discriminator_optimizer.apply_gradients(zip(
            gradients_of_discriminator, discriminator.trainable_variables))

        return gen_loss, disc_loss

    # Train the GAN
    for epoch in range(epochs):
        for batch in train_dataset:
            gen_loss, disc_loss = train_step(batch)

        # Print loss values
        print(f"Epoch: {epoch+1}/{epochs}, D Loss: {disc_loss:.4f}, G Loss:
            {gen_loss:.4f}")

        # Save generated images
        if (epoch + 1) % 10 == 0:
            generate_and_save_images(generator, epoch + 1,
                random_vector_for_generation)

    # Generate and save images
    def generate_and_save_images(model, epoch, test_input):
        predictions = model(test_input, training=False)
        fig = plt.figure(figsize=(4, 4))
        for i in range(predictions.shape[0]):
            plt.subplot(4, 4, i+1)
            plt.imshow(predictions[i, :, :, 0] * 127.5 + 127.5, cmap='gray')
            plt.axis('off')
        plt.savefig(f'image_at_epoch_{epoch:04d}.png')
        plt.show()
```

Note that this code is a simple implementation and requires more fine-tuning and a more sophisticated and robust architecture to improve the quality and efficiency of the GAN. Also, note that it is better to use diverse and high-quality data to train the GAN. To implement the GAN in MATLAB, you can use the ganNetwork and functions trainNetwork as helpers.

9.18 Deconvolution Network

A deconvolution network (DN) is a type of deep neural network used for various applications in image processing and machine vision. It is also known as an inverse convolutional network or deconvolution network.

In convolutional neural networks, convolution layers are used to extract image features. While in deconvolutional networks, the reverse of convolution is performed to obtain the original image or a higher-resolution representation of the extracted features.

This network is commonly used in tasks like image retrieval, image correction, learning high-feature representations, etc. It is also used in various fields like medical image processing, object recognition, and image signal processing. You can use the functions and layers related to deconvolution in MATLAB libraries like Deep Learning Toolbox.

To build a deconvolutional model in MATLAB, you can use functions like transposedConv2dLayer or deconvolution2dLayer, which are available in these libraries. Then create your model by adding layers and corresponding settings in order. A simple example could be as follows:

MATLAB

```
% Define the network layers
layers = [
    transposedConv2dLayer(filterSize, numFilters, 'Name', 'deconv1'),
         % Transposed convolution layer
    batchNormalizationLayer('Name', 'bn1'),
         % Batch normalization layer
    reluLayer('Name', 'relu1'),
         % ReLU activation layer
    transposedConv2dLayer(filterSize, numFilters, 'Name', 'deconv2'),
         % Transposed convolution layer
    batchNormalizationLayer('Name', 'bn2'),
         % Batch normalization layer
    reluLayer('Name', 'relu2'),
         % ReLU activation layer
    % Continue adding layers as needed
];

% Set up training options
options = trainingOptions('adam', ...
     % Optimizer
    'MaxEpochs', 50, ...
         % Maximum number of epochs
    'InitialLearnRate', 0.001);
                                                 % Initial learning
         rate

% Model training
net = trainNetwork(trainingData, layers, options);
     % Train the network
```

Python

```
import torch
from torch import nn
from torch.utils.data import DataLoader
from torchvision import datasets
from torchvision.transforms import ToTensor, Compose, Normalize

# Define the generator network
class Generator(nn.Module):
    def __init__(self, filter_size, num_filters):
        super(Generator, self).__init__()
        self.deconv1 = nn.ConvTranspose2d(num_filters, num_filters,
             filter_size, stride=2, padding=0)  # Output: 3x3
        self.bn1 = nn.BatchNorm2d(num_filters)
        self.relu1 = nn.ReLU()
```

```
        self.deconv2 = nn.ConvTranspose2d(num_filters, num_filters,
            filter_size, stride=2, padding=0)  # Output: 7x7
        self.bn2 = nn.BatchNorm2d(num_filters)
        self.relu2 = nn.ReLU()
        self.deconv3 = nn.ConvTranspose2d(num_filters, num_filters // 2,
            kernel_size=4, stride=2, padding=1)  # Output: 14x14
        self.bn3 = nn.BatchNorm2d(num_filters // 2)
        self.relu3 = nn.ReLU()
        self.deconv4 = nn.ConvTranspose2d(num_filters // 2, 1, kernel_size
            =4, stride=2, padding=1)  # Output: 28x28
        self.tanh = nn.Tanh()  # Output range [-1, 1]

    def forward(self, x):
        x = self.relu1(self.bn1(self.deconv1(x)))  # 3x3
        x = self.relu2(self.bn2(self.deconv2(x)))  # 7x7
        x = self.relu3(self.bn3(self.deconv3(x)))  # 14x14
        x = self.deconv4(x)                         # 28x28
        x = self.tanh(x)
        return x

# Define training options
num_epochs = 50
learning_rate = 0.001
batch_size = 64

# Load training data and adjust normalization range to [-1, 1]
transform = Compose([
    ToTensor(),
    Normalize((0.5,), (0.5,))
])
trainingData = datasets.MNIST(
    root="data",
    train=True,
    download=True,
    transform=transform
)
train_dataloader = DataLoader(trainingData, batch_size=batch_size, shuffle=
    True)

# Initialize the generator model
filter_size = 3
num_filters = 64
generator = Generator(filter_size, num_filters)

# Define loss function and optimizer (use MSE loss)
criterion = nn.MSELoss()  # Mean Squared Error Loss
optimizer = torch.optim.Adam(generator.parameters(), lr=learning_rate)

# Enable GPU
device = torch.device("cuda" if torch.cuda.is_available() else "cpu")
generator.to(device)
print("Using device:", device)

# Model training
for epoch in range(num_epochs):
    generator.train()
    for batch, (X, y) in enumerate(train_dataloader):
        # Input noise (batch_size, num_filters, 1, 1)
        input_noise = torch.randn(X.size(0), num_filters, 1, 1).to(device)
        X = X.to(device)  # Move real images to GPU

        output = generator(input_noise)
        loss = criterion(output, X)
```

```
        # Backpropagation and optimization
        optimizer.zero_grad()
        loss.backward()
        optimizer.step()

    print(f"Epoch [{epoch+1}/{num_epochs}], Loss: {loss.item():.4f}")
```

In this code, layers are used transposedConv2dLayer for deconvolution and batchNormalizationLayer are reluLayer used for normalization, and activation operators. You can change the number and type of layers based on your needs and the task at hand.

9.19 Liquid State Machine Network

A liquid state machine (LSM) is a type of computational model in the field of artificial neuroscience. The model is usually based on artificial intelligence theory inspired by brain function and sensorimotor systems in animal brains. In a liquid state machine, a dynamic system with fluid properties (such as liquid) is used to represent information and inputs. This system is known as a "reservoir" and, due to its dynamic nature, is able to process patterns and temporal information.

The basic idea of a fluid state machine is that input data is randomly fed into a reservoir, and then useful patterns and features are extracted by analyzing the reservoir outputs. These patterns can be used for pattern recognition problems, temporal prediction, distraction-related tasks, etc. Fluid state machines are used as an unsupervised learning algorithm for processing temporal information and are currently used in various fields including signal processing, pattern recognition, and system dynamics analysis.

The functions for fluid state machine models are not available out of the box in MATLAB and require manual implementation. Due to their complexity and primitive nature, this algorithm requires high expertise in the relevant fields. As a general guide, you can implement your own LSM model using the neural network functions available in the Deep Learning Toolbox library in MATLAB. However, note that this implementation requires research and expertise due to the complexity of the specific features of LSM.

9.20 Extreme Learning Machine

Extreme Learning Machine (ELM) is an attempt to reduce the complexity behind feedforward neural networks. It does this by creating sparse hidden layers with random connections. These types of neural networks require less computational power, but their actual performance depends heavily on the task at hand and the data.

Extreme Learning Machine (ELM) is a fast and simple type of machine learning model, mainly used for classification and prediction problems. Here is a sample code to implement ELM on training and test data:

MATLAB

```
% Example: ELM for data classification
% Definition of training data
X_train = randn(100, 10); % 100 input samples with 10 features (training data)
Y_train = randi([1, 3], 100, 1); % Category labels (training labels)

% Define test data
X_test = randn(20, 10); % 20 test input samples (test data)

% Define the number of neurons in the hidden layer
num_neurons = 50;

% Calculate weights and biases randomly
input_weights = randn(num_neurons, size(X_train, 2)); % Random input weights
bias = randn(num_neurons, 1); % Random biases

% Calculate the output of the hidden layer with the sigmoid activation function
hidden_output = sigmoid(input_weights * X_train' + bias); % Hidden layer output for training data

% Calculate output weights using pseudoinverse
output_weights = pinv(hidden_output') * Y_train; % Output weights

% Calculate model output for test data
hidden_output_test = sigmoid(input_weights * X_test' + bias); % Hidden layer output for test data
predicted_labels = round(output_weights' * hidden_output_test); % Predicted labels

% Model accuracy assessment (Note: Y_test should be defined if available)
% accuracy = sum(predicted_labels' == Y_test) / length(Y_test); % Uncomment if Y_test is available
% disp(['Model accuracy: ', num2str(accuracy * 100), '% ']); % Uncomment if Y_test is available

% Sigmoid function
function y = sigmoid(x)
    y = 1 ./ (1 + exp(-x)); % Sigmoid activation function
end
```

Python

```
import numpy as np
from numpy.linalg import pinv

# Define training data
X_train = np.random.randn(100, 10)  # 100 input samples, each with 10 features (training data)
Y_train = np.random.randint(1, 4, size=(100, 1))  # Classification labels (training labels)

# Define test data
X_test = np.random.randn(20, 10)  # 20 test input samples (test data)
```

```

# Define the number of neurons in the hidden layer
num_neurons = 50

# Randomly compute input weights and bias
input_weights = np.random.randn(X_train.shape[1], num_neurons)  #
    Random input weights
bias = np.random.randn(1, num_neurons)  # Random bias

# Compute hidden layer output (using sigmoid activation function)
hidden_output = 1 / (1 + np.exp(-(X_train @ input_weights + bias))
    )  # Hidden layer output for training data

# Compute output weights (using pseudo-inverse)
output_weights = pinv(hidden_output) @ Y_train  # Output weights

# Compute hidden layer output for test data
hidden_output_test = 1 / (1 + np.exp(-(X_test @ input_weights +
    bias)))  # Hidden layer output for test data

# Predict labels
predicted_labels = np.round(hidden_output_test @ output_weights)
    # Predicted labels

# If you have test labels, you can evaluate the model's accuracy
# In Python, if Y_test is defined, you can compute the accuracy as
    follows
# Assuming Y_test is available:
# accuracy = np.sum(predicted_labels == Y_test) / len(Y_test) *
    100
# print(f"Model accuracy: {accuracy}%")

print("Predicted labels:")
print(predicted_labels)
```

In this code, the function sigmoid defines the sigmoid activation function. ELM first randomly generates the hidden layer weights and biases, and then calculates the hidden layer output and output weights using the image matrices. Finally, the model accuracy is evaluated using the test data.

9.21 Sparse AutoEncoder Neural Network

Sparse AutoEncoder (SAE) is another type of self-encoder artificial neural network that, in some cases, can reveal some hidden group patterns in the data. The structure of the sparse autoencoder is also similar to the AE. In this type of neural network, the number of hidden layers is more than the number of input/output layer cells.

A simple example for creating a Sparse AutoEncoder with MATLAB is given below:

MATLAB

```
% Define the number of inputs and hidden neurons
inputSize = input_size; % e.g., 784 for a 28x28 image
hiddenSize = 20; % Number of hidden neurons
sparsityParam = 0.2; % Desired value
lambda = 0.6; % Regularization parameter

% Definition of cost and gradient functions
costFunction = @(params) sparseAutoencoderCost(params, inputSize, hiddenSize
    , lambda, sparsityParam, beta);

% Optimizer settings
options = optimset('MaxIter', 100, 'Display', 'iter');

% Generate random initial parameters
initialTheta = initializeParameters(hiddenSize, inputSize);

% Neural network training
[optTheta, cost] = fminlbfgs(costFunction, initialTheta, options);

% Extract weights from trained parameters
W1 = reshape(optTheta(1:hiddenSize * inputSize), hiddenSize, inputSize);

% Sparse AutoEncoder Cost Function
function [cost, grad] = sparseAutoencoderCost(theta, inputSize, hiddenSize,
    lambda, sparsityParam, beta)
    W1 = reshape(theta(1:hiddenSize * inputSize), hiddenSize, inputSize);
    b1 = theta(hiddenSize * inputSize + 1:hiddenSize * inputSize +
         hiddenSize);
    W2 = reshape(theta(hiddenSize * inputSize + hiddenSize + 1:hiddenSize *
         inputSize + hiddenSize + hiddenSize * inputSize), inputSize,
         hiddenSize);
    b2 = theta(hiddenSize * inputSize + hiddenSize + hiddenSize * inputSize
         + 1:end);

    % Neural network output calculation code
    % ...

    % Cost function calculation code
    % ...

    % Gradient calculation code
    % ...

    grad = [W1grad(:); b1grad(:); W2grad(:); b2grad(:)];
end

% Function initializes parameters
function W = initializeParameters(hiddenSize, visibleSize)
    epsilon = 0.12;
    W = rand(hiddenSize, visibleSize) * 2 * epsilon - epsilon;
end
```

Python

```
import numpy as np
from scipy.optimize import fmin_l_bfgs_b
from scipy.special import expit as sigmoid

# Parameter Initialization
input_size = 784  # Input size (e.g., size of a 28x28 image)
hidden_size = 20  # Number of hidden layer neurons
sparsity_param = 0.2  # Sparsity parameter
lambda_ = 0.6  # Regularization parameter
beta = 3.0  # Sparsity penalty coefficient

# Define the cost and gradient function
def sparse_autoencoder_cost(theta, input_size, hidden_size, lambda_,
        sparsity_param, beta, X):
    # Parameter decomposition
    W1 = theta[:hidden_size * input_size].reshape(hidden_size, input_size)
    b1 = theta[hidden_size * input_size:hidden_size * (input_size + 1)]
    W2 = theta[hidden_size * (input_size + 1):hidden_size * (input_size + 1)
         + input_size * hidden_size].reshape(input_size, hidden_size)
    b2 = theta[hidden_size * (input_size + 1) + input_size * hidden_size:]

    # Forward propagation
    m = X.shape[0]
    a1 = X  # Input layer activation
    z2 = np.dot(a1, W1.T) + b1  # Hidden layer linear combination
    a2 = sigmoid(z2)  # Hidden layer activation
    z3 = np.dot(a2, W2.T) + b2  # Output layer linear combination
    a3 = z3  # Output layer activation (linear output)

    # Components of the cost function
    # Reconstruction error
    R = (1/(2*m)) * np.sum((a3 - X)**2)

    # Weight decay (regularization)
    weight_decay = (lambda_/(2*m)) * (np.sum(W1**2) + np.sum(W2**2))

    # Sparsity penalty
    rho_hat = np.mean(a2, axis=0)  # Actual activation probability
    kl_div = sparsity_param * np.log(sparsity_param / rho_hat) + (1 -
         sparsity_param) * np.log((1 - sparsity_param) / (1 - rho_hat))
    sparsity_cost = beta * np.sum(kl_div)

    # Total cost
    cost = R + weight_decay + sparsity_cost

    # Backpropagation
    # Output layer error
    delta3 = (a3 - X) / m  # Dimension: (m, input_size)

    # Hidden layer error
    term1 = np.dot(delta3, W2)  # Dimension: (m, hidden_size)
    # term3 is the derivative of sigmoid(z2)
    term3 = a2 * (1 - a2)  # Dimension: (m, hidden_size)
    term2 = (sparsity_param - rho_hat)[None, :]  # Dimension: (1,
         hidden_size)
    delta2 = term1 * term3 + beta * term2  # Dimension: (m, hidden_size)

    # Compute gradients
    W1_grad = (np.dot(delta2.T, a1) + lambda_ * W1) / m  # Size: (
         hidden_size, input_size)
    b1_grad = np.mean(delta2, axis=0)
```

```
    W2_grad = (np.dot(delta3.T, a2) + lambda_ * W2) / m  # Size: (input_size
        , hidden_size)
    b2_grad = np.mean(delta3, axis=0)

    # Flatten gradients into a one-dimensional array
    grad = np.concatenate((W1_grad.flatten(), b1_grad, W2_grad.flatten(),
        b2_grad))

    return cost, grad

def initialize_parameters(hidden_size, visible_size):
    epsilon = 0.12
    W = np.random.rand(hidden_size, visible_size) * 2 * epsilon - epsilon
    b = np.zeros(hidden_size)
    W2 = np.random.rand(visible_size, hidden_size) * 2 * epsilon - epsilon
    b2 = np.zeros(visible_size)
    return np.concatenate((W.flatten(), b, W2.flatten(), b2))

# Example data loading
X = np.random.rand(100, input_size)  # Example input data

# Initialize parameters
initial_theta = initialize_parameters(hidden_size, input_size)

# Set optimizer options
options = {'maxiter': 100, 'disp': True}

# Use L-BFGS-B for optimization
cost_function = lambda theta: sparse_autoencoder_cost(theta, input_size,
    hidden_size, lambda_, sparsity_param, beta, X)
opt_theta, cost, info = fmin_l_bfgs_b(cost_function, x0=initial_theta, factr
    =1e12, maxiter=100)

# Extract weights
W1 = opt_theta[:hidden_size * input_size].reshape(hidden_size, input_size)
b1 = opt_theta[hidden_size * input_size:hidden_size * (input_size + 1)]
W2 = opt_theta[hidden_size * (input_size + 1):hidden_size * (input_size + 1)
    + input_size * hidden_size].reshape(input_size, hidden_size)
b2 = opt_theta[hidden_size * (input_size + 1) + input_size * hidden_size:]

print("Training completed!")
```

9.22 Deep Belief Network

A deep belief network (DBN) is a type of deep neural network that consists of a combination of different layers called the visible layer and the hidden layer. The network consists of two main types of layers: the visible layer and the hidden layer. These two layers are typically placed alternately.

DBN uses a deep belief architecture where hidden layers extract important information from the input data and then pass this information to subsequent hidden layers. This process is done incrementally, with each hidden layer acting as higher-level, harder-to-extract features.

An important feature of DBN is that each layer can be described as a Restricted Boltzmann Machine (RBM). RBM trains the observable and hidden layers independently of each other, and then these layers are combined to build the DBN.

DBNs are used as generative and transformation models to generate new data or extract latent features from input data. They have also performed well in problems such as object recognition, face recognition, machine translation, and other deep learning tasks.

Unfortunately, MATLAB libraries do not directly support implementing Deep Belief Networks. However, you can use implementations available in other languages (such as Python) and use MATLAB transformations to bring your desired results into MATLAB.

To implement DBN, libraries such as Deep Learn Toolbox (for MATLAB) or TensorFlow and Keras (for Python) are usually used. Below is a sample Python code for DBN using the Keras library. You can run this code in Python and see the result in MATLAB.

MATLAB

```
# Install Keras library
# pip install keras

from keras.models import Sequential
from keras.layers import Dense
from keras.optimizers import RMSprop
from sklearn.model_selection import train_test_split
import numpy as np

# DBN model definition
def build_dbn(input_size, hidden_layers):
    model = Sequential()

    # Add hidden layers to the model
    for units in hidden_layers:
        model.add(Dense(units, activation='sigmoid', input_dim=input_size))

    return model

# Example of using the DBN model
input_size = 784  # Example: input is a 28x28 image
hidden_layers = [500, 300, 100]  # Number of neurons in hidden layers

# Build the DBN model
dbn_model = build_dbn(input_size, hidden_layers)

# Display model architecture
dbn_model.summary()
```

9.23 Deep Convolutional Network

Deep Convolutional Network (DCN) is currently the star of neural networks. This type of neural network has convolutional cells (or pooling layers) and kernels, each serving a different purpose. The convolutional kernels process the input data, and the pooling layers simplify this task by reducing unnecessary features.

MATLAB libraries support deep convolutional neural networks (CNN), and you can use these libraries to implement deep convolutional networks. A simple example of a deep convolutional network in MATLAB is as follows:

MATLAB

```
% Determine input features
inputSize = [28 28 1]; % Input image dimensions

% Determine network architecture
layers = [
    imageInputLayer(inputSize)          % Input layer
    convolution2dLayer(3, 64, 'Padding', 'same')   % Convolution layer with 64 filters
    reluLayer() % ReLU activation function layer
    maxPooling2dLayer(2, 'Stride', 2) % Max pooling layer with stride 2
    fullyConnectedLayer(10) % Fully connected layer with 10 neurons
    softmaxLayer() % Softmax activation function layer
    classificationLayer() % Output layer for classification
];

% Define training options
options = trainingOptions('sgdm', ...          % Stochastic gradient descent with momentum
    'MaxEpochs', 10, ... % Maximum number of training epochs
    'InitialLearnRate', 0.001 ... % Initial learning rate
);

% Train the network
net = trainNetwork(trainingData, layers, options); % Train the network using training data

% Use the trained network for prediction
predictedLabels = classify(net, testData); % Predict class labels for test data

% Calculate accuracy
accuracy = sum(predictedLabels == testLabels) / numel(testLabels); % Calculate classification accuracy
disp(['Network accuracy: ' num2str(accuracy)]); % Display the accuracy
```

Python

```
import torch
import torch.nn as nn
import torch.optim as optim
from torchvision import datasets, transforms
from torch.utils.data import DataLoader

# 1. Define the neural network architecture
class CNN(nn.Module):
    def __init__(self):
        super(CNN, self).__init__()
        self.conv_layer = nn.Sequential(
            nn.Conv2d(1, 64, kernel_size=3, padding=1),  # Input channels: 1, Output channels: 64, Convolution kernel: 3x3, stride=1, padding=1
            nn.ReLU(),
            nn.MaxPool2d(kernel_size=2, stride=2)        # Pooling kernel: 2x2, stride=2
        )
        self.fc_layer = nn.Sequential(
```

```
            nn.Flatten(),
            nn.Linear(64 * 14 * 14, 10),  # Fully connected layer, input
                size: 64*14*14 (result after convolution), output size: 10
            nn.Softmax(dim=1)             # Softmax activation
        )

    def forward(self, x):
        x = self.conv_layer(x)
        x = self.fc_layer(x)
        return x

# 2. Data preprocessing
transform = transforms.Compose([
    transforms.ToTensor(),                     # Convert to tensor
    transforms.Normalize((0.1307,), (0.3081,))  # Normalize images
])

# Load dataset (assuming MNIST dataset)
train_data = datasets.MNIST(root='./data', train=True, download=True,
     transform=transform)
test_data = datasets.MNIST(root='./data', train=False, download=True,
     transform=transform)

# 3. Data loader
train_loader = DataLoader(train_data, batch_size=64, shuffle=True)
test_loader = DataLoader(test_data, batch_size=1000, shuffle=False)

# 4. Create the model
model = CNN()

# 5. Define loss function and optimizer
criterion = nn.CrossEntropyLoss()  # Use cross-entropy loss
optimizer = optim.SGD(model.parameters(), lr=0.001, momentum=0.9)  # SGD
     optimizer

# 6. Train the model
num_epochs = 10
device = torch.device("cuda" if torch.cuda.is_available() else "cpu")
model.to(device)

for epoch in range(num_epochs):
    model.train()
    for images, labels in train_loader:
        images = images.to(device)
        labels = labels.to(device)

        optimizer.zero_grad()
        outputs = model(images)
        loss = criterion(outputs, labels)
        loss.backward()
        optimizer.step()

    print(f"Epoch [{epoch+1}/{num_epochs}], Loss: {loss.item():.4f}")

# 7. Test the model
model.eval()
test_loss = 0
correct = 0

with torch.no_grad():
    for images, labels in test_loader:
        images = images.to(device)
        labels = labels.to(device)
```

```
        outputs = model(images)
        test_loss += criterion(outputs, labels).item()
        _, predicted = torch.max(outputs, 1)
        correct += (predicted == labels).sum().item()

test_loss /= len(test_loader.dataset)
accuracy = correct / len(test_loader.dataset)

print(f"Network accuracy: {accuracy:.4f}")
print(f"Test Loss: {test_loss:.4f}")
```

Here, convolution2dLayer defines the convolutional layer and maxPooling2dLayer defines the max pooling layer fullyConnectedLayer. It is a layer where all the neurons in a layer are connected and is used as a fully connected layer for classification. This is a simpler architecture, and you can modify the network to suit your task by changing the number and type of layers.

9.24 Echo State Network

The Echo State Network (ESN) is a subtype of recurrent neural network with a special training approach. Data is passed to the input and then, if supervised for multiple iterations, to the output (thus allowing for recurrent features to be involved). After this is done, only the weights between the hidden cells are updated. Of course, this type of neural network does not have significant applications.

In general, an echo state network, or ESN, is a type of recurrent neural network (RNN). In this network, a hidden layer called an echo state is created that is responsible for maintaining and transmitting information directly from the input to the output. Below is a sample MATLAB code to create an echo state network:

MATLAB

```
% Setting parameters
inputSize = 1;              % Input size
outputSize = 1;             % Output size
reservoirSize = 100;        % Echo mode size
spectralRadius = 0.9;       % Spectral radius

% Generate educational input and output
numDataPoints = 1000;
inputData = randn(1, numDataPoints);
outputData = sin(inputData);

% Construct the echo mode weight matrix
W_in = randn(reservoirSize, 1);
W_reservoir = randn(reservoirSize, reservoirSize);
W_out = randn(outputSize, reservoirSize);  % Note: This matrix should be
     learned during training

% Initialize for echo mode
x = zeros(reservoirSize, 1); % Reservoir state vector

% Echo mode training loop
for t = 2:numDataPoints
```

```
    u = inputData(t);  % Current input
    x = tanh(W_in * u + W_reservoir * x);  % Update reservoir state using
        non-linear activation
end

% Calculate output using trained echo mode (Note: Typically, W_out is
    trained using regression)
y_pred = W_out * x; % Predicted output

% Show results
figure;
plot(outputData, 'LineWidth', 2, 'DisplayName', 'Real');
hold on;
plot(y_pred, 'LineWidth', 2, 'DisplayName', 'Prediction');
xlabel('Time');
ylabel('Value');
title('Prediction using Echo State Network (ESN)');
legend('show');
grid on;
```

Python

```
import numpy as np
import matplotlib.pyplot as plt

# Setting parameters
input_size = 1              # Input size
output_size = 1             # Output size
reservoir_size = 100        # Reservoir size
spectral_radius = 0.9       # Spectral radius

# Generate educational input and output
num_data_points = 1000
input_data = np.random.randn(1, num_data_points)
output_data = np.sin(input_data)

# Construct the reservoir weight matrix (W_reservoir)
W_reservoir = np.random.randn(reservoir_size, reservoir_size)
# Adjust spectral radius
eig_vals, _ = np.linalg.eig(W_reservoir)
W_reservoir = W_reservoir * (spectral_radius / np.max(np.abs(eig_vals)))

# Input weight matrix (W_in)
W_in = np.random.randn(reservoir_size, input_size)  # Shape (100, 1)

# Output weight matrix (W_out) - This will be learned during training
W_out = np.random.randn(output_size, reservoir_size)

# Initialize reservoir state vector
x = np.zeros((reservoir_size, 1))  # Reservoir state vector (100, 1)

# Echo state training loop (Collect reservoir states)
reservoir_states = np.zeros((reservoir_size, num_data_points))
for t in range(num_data_points):
    u = input_data[:, t].reshape(-1, 1)  # Fix: Reshape u to (1, 1)
    x = np.tanh(W_in @ u + W_reservoir @ x)
    reservoir_states[:, t] = x.flatten()

# Train the output weights using linear regression (Ridge regression)
ridge_alpha = 1e-6  # Ridge regularization parameter
Y = output_data.T  # Targets (shape adjusted)
X = reservoir_states.T  # Reservoir states

```

```
W_out = Y.T @ X @ np.linalg.inv(X.T @ X + ridge_alpha * np.eye(
    reservoir_size))

# Predict using the trained ESN
predicted_output = []
reservoir_state = np.zeros((reservoir_size, 1))
for t in range(num_data_points):
    u = input_data[:, t].reshape(-1, 1)  # Fix: Reshape u to (1, 1)
    reservoir_state = np.tanh(W_in @ u + W_reservoir @ reservoir_state)
    y_pred = W_out @ reservoir_state
    predicted_output.append(y_pred.item())

predicted_output = np.array(predicted_output)

# Show results
plt.figure(figsize=(12, 6))
plt.plot(output_data.flatten(), label='Real', linewidth=2)
plt.plot(predicted_output, label='Prediction', linewidth=2)
plt.xlabel('Time')
plt.ylabel('Value')
plt.title('ESN Prediction Results')
plt.legend()
plt.grid(True)
plt.show()
```

Note that this code is a simple example of training and prediction using echo mode and may require further adjustments and modifications based on your specific task. Also, here are only the training steps and a sample of the prediction output.

9.25 Kohonen Neural Network

Kohonen Network (KN) introduces the feature of distance to the cell. This type of neural network is used for classification and tries to prepare its cells for maximum response to a particular input. When some cells are updated, their nearest neighbors are also updated accordingly. Like Support Vector Machines, this type of network is not always considered a "real" neural network.

In general, a Kohnen network is a type of self-organizing neural network that is used for self-organizing mapping of data. Here is a simple example of a Kohnen network in MATLAB:

MATLAB

```
%% Setting parameters
inputSize = 2;                      % Input size (two attributes)
mapSize = [10 10];                  % Kohonen map size
learningRate = 0.1;                 % Learning rate
epochs = 100;                       % Training epochs

%% Educational data generation
data = rand(100, inputSize);        % Generate random data

%% Initialize weights randomly
weights = rand(mapSize(1), mapSize(2), inputSize);

```

```
%% Network training
for epoch = 1:epochs
    for i = 1:size(data, 1)
        % Reshape and expand current data point to match weights dimensions
        currentData = reshape(data(i, :), 1, 1, inputSize);
        dataExpanded = repmat(currentData, [mapSize(1), mapSize(2), 1]);

        % Calculate distances (each neuron's weights to the input data)
        differences = weights - dataExpanded;
        distances = sum(differences.^2, 3);

        % Find the winning neuron (minimum distance)
        [minDistance, winnerIndex] = min(distances(:));
        [winnerRow, winnerCol] = ind2sub(mapSize, winnerIndex);

        % Calculate neighborhood function using Gaussian decay
        sigma = max(minDistance, 1e-3); % Avoid division by zero
        rowFactors = exp(-((1:mapSize(1)) - winnerRow).^2 / (2 * sigma^2))';
        colFactors = exp(-((1:mapSize(2)) - winnerCol).^2 / (2 * sigma^2));
        neighborhood = rowFactors * colFactors;

        % Expand neighborhood to 3D for element-wise multiplication
        neighborhoodExpanded = repmat(neighborhood, [1, 1, inputSize]);

        % Update weights with neighborhood consideration
        weightsDelta = learningRate * neighborhoodExpanded .* (dataExpanded
            - weights);
        weights = weights + weightsDelta;
    end
end

%% Visualize Kohonen map
figure;
scatter(data(:, 1), data(:, 2), 'filled');
hold on;
% Plot neurons
[rows, cols] = meshgrid(1:mapSize(1), 1:mapSize(2));
scatter(reshape(weights(:,:,1), [], 1), reshape(weights(:,:,2), [], 1), 'ro'
    );
title('Kohonen Map');
xlabel('Feature 1');
ylabel('Feature 2');
```

Python

```
import numpy as np
import matplotlib.pyplot as plt

# Set parameters
input_size = 2                    # Input size (two attributes)
map_size = (10, 10)               # Kohonen map size
learning_rate = 0.1               # Learning rate
epochs = 100                      # Training epochs

# Generate educational data (random data)
data = np.random.rand(100, input_size)

# Initialize weights
weights = np.random.rand(map_size[0], map_size[1], input_size)

# Network training
for epoch in range(epochs):
    for i in range(data.shape[0]):
```

```
        # Adjust the dimensions of the current data point
        current_data = data[i, :].reshape(1, 1, input_size)
        data_expanded = np.tile(current_data, (map_size[0], map_size[1], 1))

        # Calculate distances (between each neuron's weights and the input data)
        differences = weights - data_expanded
        distances = np.sum(differences**2, axis=-1)

        # Find the winning neuron (index of the minimum distance)
        winner_index = np.argmin(distances)
        winner_row, winner_col = np.unravel_index(winner_index, map_size)

        # Calculate the Gaussian neighborhood function
        sigma = max(np.min(distances), 1e-3)  # Avoid division by zero
        row_factors = np.exp(-((np.arange(map_size[0]) - winner_row)**2) / (2 * sigma**2))
        col_factors = np.exp(-((np.arange(map_size[1]) - winner_col)**2) / (2 * sigma**2))
        neighborhood = np.outer(row_factors, col_factors)

        # Expand the neighborhood weights for element-wise multiplication
        neighborhood_expanded = np.expand_dims(neighborhood, -1)
        neighborhood_expanded = np.tile(neighborhood_expanded, (1, 1, input_size))

        # Update weights
        delta = learning_rate * neighborhood_expanded * (data_expanded - weights)
        weights += delta

# Visualize the Kohonen map
plt.figure()
plt.scatter(data[:, 0], data[:, 1], s=50, facecolors='blue', edgecolors='blue', label='Data Points')
# Plot neurons
neurons = weights.reshape(-1, input_size)
plt.scatter(neurons[:, 0], neurons[:, 1], s=100, facecolors='none', edgecolors='red', marker='o', label='Neurons')
plt.title('Kohonen Map')
plt.xlabel('Feature 1')
plt.ylabel('Feature 2')
plt.legend()
plt.show()
```

9.26 Support Vector Machine

Support Vector Machine (SVM) is a machine learning algorithm for classification and regression. It is used especially for classification problems in linear and nonlinear data. The main goal of SVM is to create a plane or hyperplane in the feature space that separates the data in a way that provides the best separation between the classes.

The main feature of SVM is the use of support vectors. Support vectors are data points in the training set that help in detecting and locating hyperplanes. The hyperplane is chosen in such a way that it maximizes the minimum distance between the data points of two clusters. This distance is also known as the margin.

The SVM algorithm is divided into two types of separation: linear and non-linear. In the linear case, the hyperplane is a linear plane; in the nonlinear case, nonlinear transformations (such as kernel functions) are used to model more complex data sets. The advantages of SVM include high classification accuracy, good performance in high-dimensional spaces, and high controllability of fit.

To further explain and expand on the support vector machine, PStore has prepared a PowerPoint training set and made it available to you, which is worth reading. • SVM Support Vector Machine PowerPoint—Click.

A general structure for using SVM in MATLAB can be considered as follows. For this purpose, fitcsvm we will use the function available in MATLAB to train the SVM model. Suppose the training data consists of features X and labels y. Here is a code example:

MATLAB

```
% Educational data generation
rng(1); % Set the random number generator seed for reproducibility

% Training data generation
X_train = [mvnrnd([1, 1], eye(2), 100); mvnrnd([4, 4], eye(2), 100)]; %
    Generate training data
y_train = [-ones(100, 1); ones(100, 1)]; % Generate training labels

% Training the SVM model
svmModel = fitcsvm(X_train, y_train); % Train the SVM model

% Display decision boundary
figure;
% h = plot(svmModel); % Uncomment to plot the decision boundary
title('SVM decision boundary for two-class classification');

% Display training data
hold on;
scatter(X_train(y_train == -1, 1), X_train(y_train == -1, 2), 'g', 'filled')
    ; % Plot Class -1
scatter(X_train(y_train == 1, 1), X_train(y_train == 1, 2), 'r', 'filled');
    % Plot Class 1
legend('Decision boundary', 'Class -1', 'Class +1');
hold off;

% Test data generation
X_test = [mvnrnd([1, 1], eye(2), 20); mvnrnd([4, 4], eye(2), 20)]; %
    Generate test data
y_test_actual = [-ones(20, 1); ones(20, 1)]; % True labels for test data

% Prediction using SVM model
y_test_pred = predict(svmModel, X_test); % Predict test data labels

% Display prediction results
figure;
scatter(X_test(y_test_pred == -1, 1), X_test(y_test_pred == -1, 2), 'g', '
    filled'); % Predicted Class -1
hold on;
```

```
scatter(X_test(y_test_pred == 1, 1), X_test(y_test_pred == 1, 2), 'r', '
    filled'); % Predicted Class 1
title('SVM prediction results on test data');
legend('Predicted class -1', 'Predicted class +1');
hold off;
```

Python

```
import numpy as np
import matplotlib.pyplot as plt
from sklearn.svm import SVC
from sklearn.preprocessing import StandardScaler
from sklearn.datasets import make_classification

# Set random seed for reproducibility
np.random.seed(1)

# Generate training data
mean1 = [1, 1]
cov1 = [[1, 0], [0, 1]]
X_train1 = np.random.multivariate_normal(mean1, cov1, 100)

mean2 = [4, 4]
cov2 = [[1, 0], [0, 1]]
X_train2 = np.random.multivariate_normal(mean2, cov2, 100)

X_train = np.vstack((X_train1, X_train2))
y_train = np.concatenate((-np.ones(100), np.ones(100)))

# Train SVM model
svm_model = SVC(kernel='linear')
svm_model.fit(X_train, y_train)

# Plot decision boundary
plt.figure()
ax = plt.gca()
xlim = ax.get_xlim()
ylim = ax.get_ylim()

# Create grid points to plot decision boundary
xx = np.linspace(xlim[0], xlim[1], 30)
yy = np.linspace(ylim[0], ylim[1], 30)
YY, XX = np.meshgrid(yy, xx)
xy = np.vstack([XX.ravel(), YY.ravel()]).T
Z = svm_model.decision_function(xy).reshape(XX.shape)

# Plot decision boundary
ax.contour(XX, YY, Z, colors='k', levels=[-1, 0, 1], alpha=0.5,
           linestyles=['--', '-', '--'])
ax.set_title('SVM decision boundary for two-class classification')

# Plot training data
ax.scatter(X_train1[:, 0], X_train1[:, 1], c='g', label='Class -1',
    edgecolors='k')
ax.scatter(X_train2[:, 0], X_train2[:, 1], c='r', label='Class +1',
    edgecolors='k')
ax.legend()
ax.set_xlabel('Feature 1')
ax.set_ylabel('Feature 2')

# Generate test data
mean1_test = [1, 1]
cov1_test = [[1, 0], [0, 1]]
```

```
X_test1 = np.random.multivariate_normal(mean1_test, cov1_test, 20)

mean2_test = [4, 4]
cov2_test = [[1, 0], [0, 1]]
X_test2 = np.random.multivariate_normal(mean2_test, cov2_test, 20)

X_test = np.vstack((X_test1, X_test2))
y_test_actual = np.concatenate((-np.ones(20), np.ones(20)))

# Use the trained SVM model for prediction
y_test_pred = svm_model.predict(X_test)

# Plot prediction results
plt.figure()
plt.scatter(X_test1[:, 0], X_test1[:, 1], c='g', label='Predicted class -1',
        edgecolors='k')
plt.scatter(X_test2[:, 0], X_test2[:, 1], c='r', label='Predicted class +1',
        edgecolors='k')
plt.title('SVM prediction results on test data')
plt.legend()
plt.xlabel('Feature 1')
plt.ylabel('Feature 2')

# Display all plots
plt.show()
```

9.27 Neural Turing Machine

A neural Turing machine (NTM) is a type of neural network and a deep learning model that combines the architecture of neural networks with Turing machines. It was introduced by Alex Graves and others in 2014 and is a combination of a neural network with a memory-augmented neural network.

One of the outstanding features of NTM is that it allows the model to use an external memory to store and retrieve information. This memory operates with a structure similar to a classical Turing memory, which can have some features such as writing and reading.

In NTM, a neural network, known as a controller, is connected to an input memory and an output memory. This allows the neural network to store information in memory and use the memory to solve specific problems. These connections between the neural network and the memory are made by reading and writing content to the memory.

NTMs typically perform well on problems that require the retention and retrieval of information over time, including problems involving sequences and sequences. The model acts as a bridge between neural networks and the memory capabilities of classical Turing machines in recursive and information retrieval problems.

References

1. David S. Broomhead and David Lowe. “Multivariable functional interpolation and adaptive networks”. In: Complex Systems 2.3 (1988). A foundational paper for Radial Basis Function (RBF) networks. Relevant to Ch 9.3., pages 321–355.
2. Charles G. Broyden. “The convergence of a class of double-rank minimization algorithms 2. The new algorithm”. In: IMA Journal of Applied Mathematics 6.3 (1970). One of the key papers (along with Fletcher, Goldfarb, and Shanno) that developed the BFGS Quasi-Newton method. Relevant to Ch 1.3.4., pages 222–231.
3. Tianqi Chen and Carlos Guestrin. “XGBoost: A Scalable Tree Boosting System”. In: Proceedings of the 22nd ACM SIGKDD International Conference on Knowledge Discovery and Data Mining (2016). The paper describing the XGBoost library, which is used in Chapter 8., pages 785–794.
4. Kyunghyun Cho et al. “Learning phrase representations using RNN encoder-decoder for statistical machine translation”. In: arXiv preprint arXiv:1406.1078 (2014). This paper introduced the Gated Recurrent Unit (GRU). Relevant to Ch 9.10.
5. Corinna Cortes and Vladimir N. Vapnik. “Support-vector networks”. In: Machine Learning 20.3 (1995). The classic paper that introduced the soft-margin Support Vector Machine (SVM). Relevant to Ch 9.26., pages 273–297.
6. Ian J. Goodfellow et al. “Generative Adversarial Nets”. In: Advances in Neural Information Processing Systems (NIPS) 27 (2014). The seminal paper that introduced Generative Adversarial Networks (GANs). Relevant to Ch 9.17.
7. Alex Graves, Greg Wayne, and Ivo Danihelka. “Neural turing machines”. In: arXiv preprint arXiv:1410.5401 (2014). The original paper for Neural Turing Machines (NTMs). Relevant to Ch 9.27.
8. Kaiming He et al. “Deep Residual Learning for Image Recognition”. In: Proceedings of the IEEE Conference on Computer Vision and Pattern Recognition (CVPR) (2016). The original paper for Deep Residual Networks (ResNet). Relevant to Ch 9.11., pages 770–778.
9. Magnus R. Hestenes and Eduard Stiefel. “Methods of conjugate gradients for solving linear systems”. In: Journal of Research of the National Bureau of Standards 49.6 (1952). The original paper introducing the Conjugate Gradient (CG) method. Relevant to Ch 1.3.3 and Ch 4.4., pages 409–436.
10. Geoffrey E. Hinton, Simon Osindero, and Yee-Whye Teh. “A fast learning algorithm for deep belief nets”. In: Neural Computation 18.7 (2006). A key paper in the deep learning revival, introducing Deep Belief Networks (DBNs) and layer-wise pre-training with RBMs. Relevant to Ch 9.15 and 9.22., pages 1527–1554.
11. Geoffrey E. Hinton and Ruslan R. Salakhutdinov. “Reducing the dimensionality of data with neural networks”. In: Science 313.5786 (2006). A landmark paper demonstrating the power of Autoencoders for dimensionality reduction. Relevant to Ch 9.6., pages 504–507.

C. Zhang et al., *Practical Neural Networks in Python and MATLAB*,
https://doi.org/10.1007/978-3-032-14746-2

12. Sepp Hochreiter and Jürgen Schmidhuber. "Long short-term memory". In: Neural Computation 9.8 (1997). The original, groundbreaking paper on Long Short-Term Memory (LSTM). Relevant to Ch 8.1 and 9.9., pages 1735–1780.
13. John J. Hopfield. "Neural networks and physical systems with emergent collective computational abilities". In: Proceedings of the National Academy of Sciences (PNAS) 79.8 (1982). The classic paper that introduced the Hopfield Network. Relevant to Ch 9.13., pages 2554–2558.
14. Guang-Bin Huang, Qin-Yu Zhu, and Chee-Kheong Siew. "Extreme learning machine: theory and applications". In: Neurocomputing 70.1-3 (2006). The key paper introducing the Extreme Learning Machine (ELM). Relevant to Ch 9.20., pages 489–501.
15. Herbert Jaeger and Harald Haas. "Harnessing nonlinearity: Predicting chaotic systems and saving energy in wireless communication". In: Science 304.5667 (2004). A key paper on Echo State Networks (ESNs), which are closely related to Liquid State Machines (LSMs). Relevant to Ch 9.19 and 9.24., pages 78–80.
16. Simon J. Julier and Jeffrey K. Uhlmann. "Unscented filtering and nonlinear estimation". In: Proceedings of the IEEE 92.3 (2004). A comprehensive overview of the Unscented Kalman Filter (UKF) by its creators. Relevant to Chapter 7., pages 401–422.
17. James Kennedy and Russell Eberhart. "Particle swarm optimization". In: Proceedings of ICNN'95-International Conference on Neural Networks 4 (1995). The original paper introducing Particle Swarm Optimization (PSO). Relevant to Chapter 6., pages 1942–1948.
18. Diederik P. Kingma and Jimmy Ba. "Adam: A method for stochastic optimization". In: arXiv preprint arXiv:1412.6980 (2014). Introduced the Adam optimizer, the default choice for training most deep neural networks. Highly relevant to Ch 1.
19. Diederik P. Kingma and Max Welling. "Auto-encoding variational bayes". In: arXiv preprint arXiv:1312.6114 (2013). The original paper for Variational Autoencoders (VAEs). Relevant to Ch 9.8.
20. Teuvo Kohonen. "Self-organized formation of topologically correct feature maps". In: Biological Cybernetics 43.1 (1982). The seminal paper on Self-Organizing Maps (SOMs), also known as Kohonen Networks. Relevant to Ch 9.25., pages 59–69.
21. Yann LeCun et al. "Gradient-based learning applied to document recognition". In: Proceedings of the IEEE 86.11 (1998). The foundational paper on Convolutional Neural Networks (LeNet-5). Relevant to Ch 8 and 9.23., pages 2278–2324.
22. Donald W. Marquardt. "An algorithm for least-squares estimation of nonlinear parameters". In: Journal of the Society for Industrial and Applied Mathematics 11.2 (1963). The foundational paper for the Levenberg-Marquardt (LM) algorithm. Relevant to Ch 1.3.5 and Ch 4.3., pages 431–441.
23. Frank Rosenblatt. "The perceptron: a probabilistic model for information storage and organization in the brain". In: Psychological Review 65.6 (1957). The original paper introducing the Perceptron. Relevant to Ch 2.4 and 9.1., page 386.
24. David E. Rumelhart, Geoffrey E. Hinton, and Ronald J. Williams. "Learning representations by back-propagating errors". In: Nature 323.6088 (1986). The seminal paper that popularized the backpropagation algorithm for training MLPs. Relevant to Chapter 2., pages 533–536.
25. Sean J. Taylor and Benjamin Letham. "Prophet: forecasting at scale". In: The American Statistician 72.1 (2018). The paper describing the Prophet forecasting model, which is used in Chapter 8., pages 37–45.
26. Pascal Vincent et al. "Extracting and composing robust features with denoising autoencoders". In: Proceedings of the 25th International Conference on Machine Learning (ICML) (2008). The foundational paper for Denoising Autoencoders (DAEs). Relevant to Ch 9.7., pages 1096–1103.
27. Christopher M. Bishop. Pattern Recognition and Machine Learning. A classic textbook providing a thorough mathematical foundation for many machine learning concepts, including neural networks and SVMs. Springer, 2006.
28. François Chollet. Deep Learning with Python. 2nd. The authoritative guide to Keras, written by its creator. Essential for practical deep learning in Python. Manning Publications, 2021.

29. Phil Galli. Deep Learning with MATLAB: Build and deploy deep learning models using MATLAB and its toolboxes. A modern, practical guide focused on MATLAB's Deep Learning Toolbox, relevant to the book's dual-language approach. Packt Publishing, 2020.
30. Aurélien Géron. Hands-on Machine Learning with Scikit-Learn, Keras, and TensorFlow. 3rd. An extremely popular practical guide for implementing machine learning in Python, directly relevant to the book's Python examples. O'Reilly Media, 2022.
31. David E. Goldberg. Genetic Algorithms in Search, Optimization, and Machine Learning. The foundational book on Genetic Algorithms (GA), relevant to Chapter 5. Addison-Wesley, 1989.
32. Ian Goodfellow, Yoshua Bengio, and Aaron Courville. Deep Learning. The definitive, comprehensive textbook on modern deep learning. Covers fundamentals, CNNs, RNNs, and more. MIT Press, 2016. URL: http://www.deeplearningbook.org.
33. Martin T. Hagan, Howard B. Demuth, and Mark H. Beale. Neural Network Design. The classic textbook for neural networks using MATLAB, foundational for the MATLAB-specific content. PWS Publishing Co., 1996.
34. Trevor Hastie, Robert Tibshirani, and Jerome Friedman. The Elements of Statistical Learning: Data Mining, Inference, and Prediction. 2nd. A foundational text in statistical learning, providing deep insights into SVMs, optimization, and model evaluation. Springer, 2009.
35. Simon S. Haykin. Neural Networks and Learning Machines. 3rd. A comprehensive and classic textbook covering a wide array of neural network architectures, including those in Chapter 9. Prentice Hall, 2009.
36. Wes McKinney. Python for Data Analysis: Data Wrangling with Pandas, NumPy, and IPython. 2nd. The essential guide to the Python data stack (NumPy, pandas), a prerequisite for all Python examples in the book. O'Reilly Media, 2017.
37. Jorge Nocedal and Stephen J. Wright. Numerical Optimization. 2nd. The standard graduate-level textbook on optimization, covering Gradient Descent, Newton's method, CG, and Quasi Newton (BFGS/L-BFGS). Relevant to Ch 1 4. Springer, 2006.
38. Vladimir N. Vapnik. Statistical Learning Theory. The foundational work on statistical learning and Support Vector Machines (SVMs) by their inventor. Relevant to Ch 9.26. Wiley, 1998.

The manufacturer's authorised representative in the EU is Springer Nature Customer Service Centre GmbH, Europaplatz 3, 69115 Heidelberg, Germany. If you have any concerns regarding our products, please contact ProductSafety@springernature.com

Printed and bound by CPI Group (UK) Ltd, Croydon, CR0 4YY
07/07/2026
02160926-0003